ROUTLEDGE LIBRARY EDITIONS:
ENVIRONMENTAL POLICY

Volume 3

DILEMMAS

DILEMMAS

Coping with Environmental Problems

GRAHAM BENNETT

LONDON AND NEW YORK

First published in 1992 by Earthscan Publications Ltd

This edition first published in 2019
by Routledge
2 Park Square, Milton Park, Abingdon, Oxon OX14 4RN

and by Routledge
52 Vanderbilt Avenue, New York, NY 10017

Routledge is an imprint of the Taylor & Francis Group, an informa business

British Library Cataloguing in Publication Data
A catalogue record for this book is available from the British Library

ISBN: 978-0-367-18894-8 (Set)
ISBN: 978-0-429-27423-7 (Set) (ebk)
ISBN: 978-0-367-19322-5 (Volume 3) (hbk)
ISBN: 978-0-429-20175-2 (Volume 3) (ebk)

Publisher's Note
The publisher has gone to great lengths to ensure the quality of this reprint but points out that some imperfections in the original copies may be apparent.

Disclaimer
The publisher has made every effort to trace copyright holders and would welcome correspondence from those they have been unable to trace.

DILEMMAS

Coping with environmental problems

Graham Bennett

First published 1992 by
Earthscan Publications Ltd
3 Endsleigh Street, London WC1H 0DD

British Library Cataloguing in Publication Data
Bennett, Graham
Dilemmas: coping with environmental problems
I. Title
333.7

ISBN 1-85383-021-6

Typeset by Bookman Ltd, Bristol
Printed and bound by Clays Ltd, St Ives plc

Earthscan Publications Ltd is an editorially independent subsidiary of Kogan Page Limited, publishing in association with the International Institute for Environment and Development and the World Wide Fund for Nature (UK).

CONTENTS

For

my parents
who started me out along this road

Odilia
my travelling companion

and

Hanneke and Maaike
who have such a long journey ahead of them

ACKNOWLEDGEMENTS

There can be few tasks as frustrating as attempting to convey the extent to which I am indebted to all those who have willingly provided information and advice during the preparation of this book. Worse, of the hundreds of people I consulted, it is only feasible to acknowledge a small proportion by name. This does not imply any lack of gratitude to those whose contribution is not specifically acknowledged, only a recognition of the limits on space. I trust that they will accept both my thanks and my apologies.

In researching the Tellico Dam case, David Etnier of the University of Tennessee took a genuine interest in my queries and generously provided far more information than I had any right to expect. As the discoverer of the snail darter, Dr Etnier's role in the issue was crucial and my analysis benefited greatly from his constructive advice. He also kindly supplied some of the photographs which illustrate the case. The Tennessee Valley Authority itself, although subjected to much criticism of its role in the issue, responded to my request for assistance in a positive way and willingly answered all my questions. It also gave permission to use some of the material from the environmental impact statement. Further assistance, documentation and illustrations were provided by James D. Williams of the US Fish and Wildlife Service.

Information on developments in the Rhine issue was kindly provided by Henk van Hoorn of the Department of International Water Policy at the Dutch Public Works Department and a delegate to the International Rhine Commission. The legal complexities of the series of court actions which, at the time of writing, had spanned 17 years were unravelled by Jan van Dunné, Professor of Civil and Commercial Law at the Institute of Environmental Damage of the Erasmus University, Rotterdam.

An authoritative account of the trials and tribulations involved in developing an EC policy on the control of acid emissions was provided by two of the participants, Peter Stief-Tauch and Robert Donkers of the Directorate-General for the Environment, Nuclear Safety and Civil Protection of the European Commission. Their willingness to discuss the sensitive political issues involved in the case is greatly appreciated and made an invaluable contribution to my understanding of the events. I am also grateful to my colleague Nigel Haigh, Director of the London office of the Institute for European Environmental Policy, for commenting on an early draft of the chapter. An earlier version of the chapter was presented as a paper under the title *The EC Large Combustion Plant Directive* to the conference "Envrisk '88" in Como, Italy, in May 1988.

My research into the bowhead whaling case was greatly assisted by three renowned experts on the issue. Ray Gambell, Secretary of the International Whaling Commission, gave me access to his office's comprehensive documentation and suggested other sources of information. Sidney Holt, an international authority on whales and deeply involved for many years in efforts to protect whale populations, provided both an expert view on the bowhead issue and much additional material. Further insights and information were provided by Kees Lankester, a member of the Dutch delegation to the International Whaling Commission, who kindly gave up a Sunday morning to discuss the issue with me.

Expert advice on some of the technical issues involved in the Bhopal case was willingly given by Marcel Flipsen of the Dutch Labour Inspectorate. The agency also assisted in tracing some of the specialist documentation dealing with the accident.

During my visit to Italy to study the Magra valley case, I stayed for four days with Luigi Biso, the president of the local branch of Italia Nostra and one of the most prominent figures in the issue. Apart from providing me with considerable information and documentation on the issue, he also introduced me to many of the key actors and made my stay in Bocca di Magra a memorable occasion. No words can express my appreciation of the hospitality offered by Dr Biso and his family. I must also express my thanks to the shipyard involved in the case, Intermarine, and particularly to Giuliano Thermes and Andrea Grassi who made time available to give their own perspective on the issue. Generous advice, information and assistance was also provided by Roberto Lasagna of WWF Italy, Sauro Baruzzi, the President of the commune of La Spezia during the period 1980–85, Giuseppe Sansoni and Piero Sacchetti of the Massa Public Health Unit, Giovanni Raggi of the the University of Pisa and Elena Tartagione. Hans Deichmann was most helpful in preparing my visit to Bocca di Magra. I am also indebted to Giovanne Vrijenhoek-Nave for her speedy and meticulous translations of many of the Italian documents relating to the case.

It was especially rewarding for me that Tim O'Riordan, Professor of Environmental Sciences at the University of East Anglia, should agree to write a foreword to the book. He was the most important formative influence during my years at East Anglia and is a valued friend and intellectual mentor. His comments on my treatment of the main issues raised in the book were pertinent, as always, and I hope that the results do not unduly betray his counselling over the past 17 years.

To my editor, Neil Middleton, I should apologise for failing so conspicuously to produce a final manuscript anywhere near the original deadline. His inexhaustible patience and persistent encouragement are essential qualities for an editor, but they were still far more than I deserved.

It cannot go unrecorded that my two daughters, Hanneke and Maaike, were born during the writing of the book and have seen less of their father than either they or their father anticipated. It would be some compensation for all three of us if they were one day to appreciate the significance of the issues raised by the book.

But my greatest debt is to my wife, Odilia. This book could not have been written if she had not provided full support and displayed exceptional tolerance during the inordinately long period from original proposal to final manuscript.

Graham Bennett
Beek-Ubbergen, the Netherlands
April 1991

PREFACE

by Professor Timothy O'Riordan

In his marvellous series of lectures entitled *On Reconciling Man with his Environment*, Lord Ashby cogently observed that the environmental idea advanced through conflict, dilemma and reconstruction. Such struggles, he considered, are immensely useful, for they provoke a continuing debate about moral choice: choice between hard and soft values, choice between indulgence in the present and consideration for the future. He added that such dilemmas oblige people to strike a balance between what can be quantified and caring for what cannot be quantified. Thus the very framework that requires choice creates the evolutionary drive to implant nature further into the human conscience. It forever reveals the fundamental incompatibility between a still colonizing and expanding species and an earth in equilibrium.

There is little doubt that such dilemmas reveal the inconsistencies in the human condition, in the inequalities and injustices of human values and its governments, and in the schizophrenia that resides in all of us when contemplating our natural role. Equally it is important to examine such cases with great care, to discover why a "problem" is interpreted in so many contrasting ways. Take, for example, the shortage of water in Southern California. For some the problem is simply a matter of a deficit caused by an imbalance of supply and demand, a matter of engineering. For others it is the consequence of an historical legal arrangement that encourages some users to squander water at a highly subsidized price, while others run short. For others still it is an outcome of the clash of powerful interests who can carve up a common resource to maintain their hold over the order of things, but who will give way in gesture politics in order to placate the environmental fringe.

A good environmental case study should tell us a lot about history, economics, the law and raw political power along with its influence over government. It should also inform us about the processes that run the earth, about the earth and natural sciences, which set the outer limits on human actions, so long as those limits can be identified and accepted by common consent. Sadly this is rarely the case. This is why human strengths and weaknesses, the pattern of civil rights and social justice, and the legacy of past commitments and political indebtedness play a more significant part in explaining why such dilemmas exist in the first place and how they can be resolved.

There are precious few case studies of this type in the literature. Those that are written up are mostly dated, containing much that is no longer a matter of consequence. It is inevitable that some of the lessons of failure, as well as of success, are being learned. So this collection of essays by Graham Bennett is particularly to be welcomed. They will provide a valuable addition to the reading material in environmental politics courses. The skill with which he writes will also be welcome. The text reads easily, with the nuances of event, personality, institution and ill-resolved dilemmas being clear for all to follow. The international variety as well as the topic mix are also illuminating, for all these dilemmas could, and in many cases have, appeared in some form or another all over the world.

What is of special interest in this collection is that in every case the environmental interest, in its widest sense, has lost. Of course it is true that there have been important gains in, say, the reduction of salt in the Rhine, or of sulphur dioxide in Western European total emissions. But the sobering conclusion about most environmental struggles is that they accommodate to the status quo rather than change it. The real, and formidable, structure of power that runs human societies is, for the most part, unscathed. In the Tellico Dam case, the dam was finally constructed, even though it was essentially un-needed and uneconomic. One cannot imagine a story ending in which the dam remains incomplete – a monument to human avarice and miscalculation of the balance between the quantifiable and the unquantifiable. In the Rhine brine example, the protracted delay in obtaining a satisfactory outcome allowed plenty of salt to flush through the river, made many Dutch horticulturalists worse off, and feathered the nests of numerous engineering and economic consultants. The final outcome favoured the interests of an outmoded industry as much as it satisfied the legitimate demands of the market gardeners and environmentalists.

The acidification issue is by no means ended with the passage of the Large Combustion Plant Directive. No one seriously thought it would be, so the agreement that produced half an SO_2 removal loaf allowed the politicians to agonize over other things such as North Sea seal deaths and the introduction of the catalytic converter. Seals still die by the way, of causes unknown and unnatural, and urban air pollution worsens by the year. Everywhere the politicians tamper but fail to resolve.

Nowhere is this more evident that in the case of Bophal. If that event had occurred in California or Teesside there would have been a most unholy row. As it was in India, Union Carbide has still not paid out any substantial compensation, arguing spuriously that the real cause of mischief was sabotage not negligence. This is a monstrous smokescreen that allows mendacious people in many arms of the company, international finance, and government in India and the US to get off the hook, while tens of thousands of innocent people continue to suffer terrible pain and shattered lives.

Even the comical Magra valley case is a farcical tragedy. The naval boats got out (that was inevitable), again via a process that financed many an engineering consultancy, and industrial development in the valley is by no means ended. The valley remains vulnerable, and after a decorous period of time while memories lengthen and the old guard get long in the tooth, some cosmetically environmentally friendly industrial renaissance will doubtless occur.

Lord Ashby implied that the agony of each dilemma resulted in the incremental shift in the fulcrum between humanity and the natural world in favour of the latter. One wonders. This book tells us more about the triumphal escapism of those in the environmental wrong, than about the environmental righteous. The lessons we are learning concern how to roll with the punch, rather than to reconstruct the rules.

Timothy O'Riordan
School of Environmental Sciences
University of East Anglia
1991

CHAPTER ONE

INTRODUCTION

The dilemmas of environmental management

You are chairman of a public works authority with responsibility for supervising a major river development project. With the project nearly completed a problem arises. The sole population of a previously unknown species of fish is discovered in the river. If construction work continues the habitat of the fish will be destroyed and the species will become extinct. Yet, unless the project is completed, there will be no return on the huge investment. What do you do?

Again, you are the political head of a community located on the upper reaches of a major international waterway. The economy of the community is largely dependent on a local mine, but the mine is only economically viable if it can dispose of the large quantities of spoil at low cost. For the past 50 years this has been done by dumping the material in the river. But due to a steady increase in both the quantity of waste produced by the mine and other polluting discharges along the river, the quality of the water has deteriorated to the point where it is often unsuitable for use by the riparian states downstream. What do you do?

Or again, you are the prime minister of a highly industrialized country. Recent scientific evidence suggests that the atmospheric emissions from large industrial installations in your country are carried by the prevailing winds over long distances and cause severe environmental damage in other countries. Because of its favourable location, your own country receives little airborne pollution from foreign sources. Strict international controls on these types of emissions are proposed. Your country, as one of the major sources of the emissions, would have to make substantial investments in the necessary pollution abatement measures. But most of the environmental improvements would be seen in other countries; the benefits to your own country would be minimal. What do you do?

Once again, you are the head of an indigenous culture which has depended for thousands of years on the sustainable yield of a large mammal. But foreign, technologically advanced hunters have also discovered the animal and have reduced its numbers to the point where it is threatened with extinction. The predicament of the creature is now widely recognized and an international call is made to protect the remaining specimens. To continue hunting the animal would carry the risk that it would quickly become extinct; to refrain from further hunting would be a denial of the essence of your culture and inevitably lead to its degeneration. What do you do?

Yet again, you are the state governor in a Third World country scarred by deprivation. A multinational company proposes to build a chemical plant on the outskirts of the regional capital. The plant would create hundreds of jobs and make an important contribution to the economic development of the area. One of the products to be manufactured in the plant involves a hazardous process. The company assures you that the plant will be designed and operated to ensure that no undue risk will arise, but your rudimentary technical inspectorates are not qualified to make an independent judgment and do not have sufficient resources to monitor the plant effectively. What do you do?

Finally, you are mayor of a small community. The main town is built at the mouth of a picturesque river valley which harbours many rare aquatic ecosystems. One of the more important local enterprises is a shipyard specializing in the production of small pleasure craft. It is located just upstream of a low bridge which carries the main coastal road over the lower stretch of the river. The shipyard unexpectedly announces that it has won a contract to build a number of naval vessels with the possibility of more orders to come and the prospect of important economic benefits to the local economy. The vessels, however, will be too tall to pass under the bridge to the sea, and the company requests permission to replace one of the fixed spans with a lifting section. But a local environmental group points out that this modification would encourage the expansion of the shipyard, the construction of new shipyards and, by also allowing larger ships to pass upstream, would open up the river valley to further industrial development. What do you do?

This is a book about the hard reality of environmental management. It describes how each of these dilemmas arose and shows what happens in practice when we are confronted with these kinds of problems. But the concerns raised by the problems extend far beyond the scope of the six case studies, for they go to the heart of environmental management. The central themes of the book therefore concern the nature of environmental problems, the way such problems arise, our ability to anticipate the unintended and undesirable environmental effects of our actions and the way in which we respond to environmental impacts once they arise. And ultimately it is a book about failure – the failure to understand properly the issues posed by environmental problems and the failure to develop adequate mechanisms for preventing and controlling environmental impacts.

The case studies that make up the main part of the book graphically illustrate the forces that impede our efforts at environmental control. Each case is a notorious example of an intractable environmental problem: the construction of the Tellico Dam; the disposal of waste salt in the Rhine; the acid emissions control policy of the European Community;

the hunting of the bowhead whale; the disaster in Bhopal and the protection of the Magra valley. Some earned greater notoriety than others, but all were resistant to easy resolution. They have been selected not because they are representative of environmental problems in general – a sample of six is inadequate for that purpose – but because they are unusually powerful illustrations of the difficulties involved in controlling the environmental impacts of our actions.

A contributory reason for selecting these issues concerns the insights they provide into the special nature of the choices that environmental problems pose. In considering each of the cases, it is important to bear in mind the fundamental nature of the respective problems, for it was the intrinsic structure of each which guaranteed that no straightforward remedy could be devised. It is often assumed that for every problem there is a solution – that somewhere among the various alternative courses of action lies the option which offers a painless way of remedying the situation. In the case of environmental problems this assumption is a delusion. Environmental problems do not offer solutions, they present dilemmas: we are confronted with choices between alternatives, all of which carry their own disadvantages. In choosing the benefits of one course of action we are compelled to forego the advantages of alternative options. The introductory examples show, for instance, that we cannot choose to have both the public works project and the fish, or both an effective international emissions control agreement and a high national environmental return on the pollution control investments in each country. In making these choices we not only have to choose between conflicting objectives, but at the same time we have to allocate to the interested parties the gains and losses which are associated with that choice. To be sure, it may in some instances be feasible to devise compromises that soften the blow by optimizing the trade-offs between different objectives – less pollution and higher abatement costs for example – but the essential nature of the dilemma remains.

The kinds of dilemmas presented by environmental problems can be stated quite succinctly. First and foremost are ethical dilemmas: do individual organisms, communities or species have a right to exist? If we take the "biocentrist" view that humans are an integral part of nature, we might conclude that evolutionary principles require the maximization of ecological diversity. On the other hand, our observations of nature do not suggest that all natural life should be regarded as sacred; certainly, the behaviour of other species does not suggest that they operate on the principle of causing the least possible damage to their environment. If a "dualist" approach is adopted, in which people observe the natural world from a separate and objective perspective, our moral code may require some form of respect for the natural world. But respect is a subjective notion which involves questions of degree: how much of the tropical rainforest may we destroy? A bioethic, if it is to have any meaning at all, should surely be axiomatic, not contingent on the exigencies of any particular circumstances. As we shall see, in the absence of immutable principles, utility will always tend to displace moral justice.

Second are efficiency dilemmas: how much environmental damage is acceptable? In contrast to ethical dilemmas, questions of degree are central to efficiency choices. If we accept that increasing the quality of our existence justifies actions which damage the environment beyond the absolute minimum level that would be necessary only for our survival, we are confronted with the need to trade off the utility of our primary objectives against the environmental damage caused by their secondary effects, that is, to deduce a proper balance between utility and environmental damage. To state the problem in rational terms, the task is to determine the optimum level of environmental damage. But although economic theory provides an array of conceptual tools for defining this optimum level, they are not sufficiently operational to be able, in practice, to determine a proper balance between the conflicting objectives. Thus the sole economic indicator of utility is purchasing power, which implies that the preferences of the rich carry more weight than those of the poor (in fact the distribution of wealth is not regarded as a matter over which economics has any jurisdiction). Neither does neoclassical economics recognize value judgments that are not expressed in a market, yet many "environmental goods", the natural beauty of a river valley, for example, cannot be marketed. We are therefore faced with the challenge of deciding the acceptable level of environmental damage without a comprehensive calculus of efficiency. The scope which this offers for rather more opportunistic approaches to deciding the proper level of environmental damage is a persistent theme in the case studies.

Third are equity dilemmas: who benefits from the activities that cause environmental impacts, and who suffers the damage? Who pays for the costs of abatement measures, and who benefits from the environmental improvements? In a perfect world, the distribution of the benefits generated by an enterprise would be congruent with the distribution of the associated costs of the activity; in other words, only those who enjoyed the benefits would bear the costs (and ideally, other things being equal, in direct proportion to the benefits enjoyed). In the imperfect world, however, we find that the environmental costs of various enterprises are borne by groups who receive either no benefit at all, or only a minimal benefit from the respective activities. These inequities are distributed not only between different social groupings (intersocial equity), but also over space (interspatial equity) and over time (intergenerational equity). Moreover, not all people in each group will be affected in the same way; those who "consume" environmental resources – such as natural beauty or the assimilative capacity of a river – will suffer disproportionately. Yet we have no formal mechanism for dealing with environmental equity, with the result that in practice the potential victims of inequity become vulnerable to the exercise of interest group power rather than social justice.

Fourth are liberty dilemmas: to what extent should we restrict individual freedom in the interests of environmental protection? In any society there are conflicts between the perceived interests of the individual and the collective interests of the community. Each of us is well aware of the temptation to pursue our own interests at the expense of others. The obvious result is a redistribution of gains and losses: my gain is someone else's loss. But another possible result is less obvious: large numbers of individuals or groups acting in their own interests may, in aggregate, produce consequences that nobody, not even the actors themselves, may desire. The result is not only a redistribution of gains and losses but a net loss for the community as a whole. Environmental damage caused by the aggregate effect of a multitude of atomistic decisions is a good example of such a collective loss. In many cases, of course, environmentally friendly behaviour is suitably guided by appropriate incentives (or disincentives): governments lay down emission standards, together with the threat of a fine for non-compliance, to limit the quantities of polluting substances which industrial installations may discharge into the atmosphere. But many other cases are either not amenable to social control or do not justify the expense and effort required to establish and enforce a control regime – or, as will be seen in the case studies, certain interests strongly resist subordinating their private interest to the collective interest.

Fifth are uncertainty dilemmas: how do we make choices on the basis of inadequate information? There are few actions where the entire sequence of consequences can be predicted with certainty. We cannot completely control all the circumstances which may influence an event: a power failure at a crucial moment, for example; or the conditions under which the events occur, such as the weather; or the long and complex chain of reactions which follow an event.

In fact it may be far from certain whether a problem is likely to arise or has already arisen. We may also be undecided on the evidence we regard as necessary in order to justify corrective action: do we wait until research or an accident shows unequivocally that a problem exists, or do we take precautionary action on the basis of an indication that something is amiss? It is a question of where the burden of proof should lie: innocent until proved guilty or guilty until proved innocent? Either course of action has its disadvantages. If we decline to take action until indisputable evidence is available that a problem exists, we risk unnecessary damage if it transpires that there is a problem; if we take precautionary measures to forestall potential threats, we risk incurring unnecessary costs if it later transpires that either there was in fact no problem or that the measures taken were inappropriate. Further, our judgment of whether particular environmental problems are serious enough to merit corrective action is determined not simply by rational analysis but by our perception of the issue, by our values and beliefs: there is no objective measure of what constitutes an acceptable level of environmental risk. And this is particularly difficult when an activity promises substantial benefits but carries the small risk of a catastrophic accident, as was the case in Bhopal.

Last, but not least, are evaluation dilemmas: how should we compare incommensurate values? In order to make informed choices between different objectives, between industrial development and environmental protection, for example, it is necessary to ascribe values to the various alternatives. We therefore need some means of evaluating the consequences of our actions. The key to evaluation is the application of a common rating system. If all costs and benefits can be represented by commensurate units, they can be added and compared and a rational choice can be made. But no matter how objective a rating system may appear, all ratings are value judgments. And if there is no objective measure of personal values, they can be neither quantified nor aggregated into group preferences. Attempts to deal with this dilemma generally involve devising mechanisms for ascribing numerical surrogates to different values, and the most widespread and readily grasped surrogate is money. However, prices cannot provide information on consumer preferences where there is no market, and for many environmental "goods" there is no market. The temptation is therefore to find quantitative surrogates for the intangible values such as natural beauty and biological diversity so that these can compete on an equal footing with "hard" economic values. The difficulty in applying techniques such as cost-benefit analysis is the impression they convey that the restricted conception of welfare which they represent is an accurate representation of environmental values. That attempting the rigorous evaluation of different options is the exception rather than the rule in environmental management is apparent from the case studies; that many of the attempts, when made, are patently superficial is even more striking.

A final reason for selecting these case studies is the way in which they illustrate how individuals, interest groups and political institutions respond to environmental threats of various kinds. Thus we see, for example, failures to invest in rigorous anticipatory mechanisms, so that in each case the initiation of the activity which caused the environmental impact preceded any proper assessment of its likely effects.

We see a legally established and immutable ethical principle degraded to a policy objective by the very first project which put the principle to the test. And we subsequently find that, when a policy review leads to a confirmation of the objective, a one-off political coalition can still succeed in exempting the project from that decision.

We see minority interests which press their conspicuous claims by applying concentrated pressure on decision-making institutions, effectively ensuring that their claims become impregnable in the political and social consciousness. These consciousness-manipulating efforts can succeed to the point where analyses showing a claim to have negative economic benefits lose much of their persuasive force. At the same time we can observe the exceptional vulnerability of formal authorization procedures to the application of judicious interest group pressure and the astute use of procedural opportunities, such as the exercise of veto powers, to advance or defend a claim.

We see a remarkable lack of determination to reduce the uncertainties associated with certain courses of action, particularly where reducing those uncertainties may act to the disadvantage of prominent constituencies. Moreover, we can discern substantial discrepancies between the potential environmental threats of an activity and the resources devoted to assessing those threats and evaluating their costs.

We see the propensity of separate interests to negotiate deals which offer mutual advantage (or a minimum of disadvantage) rather than collectively optimal compromises. We can observe that the pressures to strike such deals appear to bear an inverse relationship to the openness of the decision-making process and the opportunities for the interests concerned to demonstrate the propriety of their agreements.

Finally, we see that in none of the six cases did the trade-offs between economic objectives and environmental protection turn out unequivocally to favour the environment. And this is despite the fact that in virtually all the cases the environmental dimension of the issue was given unusual prominence. Although there is no *a priori* reason to conclude that environmental protection should have been given predominance in these cases, the unusual importance of the environment or species under threat, and the marginal nature of the economic interests often involved, clearly suggests a structural imbalance in the way we resolve these kinds of issues. Specifically, the case studies show that economic constituences are demonstrably better equipped to cope with conflicts of interests than are environmental constituencies.

The general conclusion to be drawn from these observations is that the outcomes of the environmental control decisions in the six cases tended to be governed by the way private interest groups wielded their power and influence. Power is always exercised on behalf of certain interests, and the extent to which power is concentrated in society, the distribution of that power and the relationships between those who hold the power will largely define the processes which shape environmental problems and the way in which society responds to those problems. What the case studies show is that, in the absence of representation by a powerful interest, environmental protection is consistently subordinated to the pursuit of private, short-term objectives.

It is to counter this tendency, and in direct response to the broad social concern about environmental degradation, that an extensive array of environmental management tools have been developed since the 1970s. Instruments such as environmental impact assessment, the polluter pays principle, risk analysis, environmental auditing and strict liability are designed specifically to articulate the environmental interest in the decision-making process. The key question that we now face is whether these instruments can be elaborated and vested with a powerful enough constituency to facilitate a fundamental change in our relationship with the environment, or whether their substantive shortcomings and lack of political credibility ultimately ensure that they act only on the margins of the issue. The evidence of the case studies is not encouraging in this respect; they show that these kinds of instruments tend to be of greater procedural than substantive significance and that their contributions to the decision-making process can be readily disregarded when a moderately powerful coalition is assembled in support of a conspicuous interest. It is here that the real challenge for environmental management lies: whether it can succeed not only in articulating environmental imperatives but in vesting those imperatives with sufficient weight to provide a countervailing force in resolving environmental dilemmas.

CHAPTER TWO

A SHAGGY FISH STORY

The snail darter and the Tellico Dam

On Sunday, 12 August 1973, David Etnier was collecting fish in the shallow shoals of Coytee Spring on the lower reaches of the Little Tennessee River. To Etnier, a zoologist from the University of Tennessee, the trip was little different from the many previous visits he had made to the area as part of his research into the fish fauna of this extensive river basin. Equipped with face mask and snorkel, he entered the clean, fast-flowing waters and dived down towards the river bed. What happened next has remained clear in his memory ever since:

> The first fish I saw was what I thought to be a rather strange-looking and slender sculpin. I poked at it several times to obtain alternate perspectives and still wasn't convinced that it was a sculpin. I was able to catch the fish with my hands, and upon standing up and removing my face mask was astonished to see a darter species that I immediately realized no-one had ever seen before.

David Etnier was right. His first catch of the day was indeed an unknown type of darter. Although interesting to the zoologist, the discovery could hardly be described as headline news. Novel plant and animal species are recorded virtually every day somewhere in the world; in the normal course of events the find would merit a paper in an obscure scientific journal and would then be promptly forgotten. This time things were to work out differently. Very differently, in fact, for the fish's tranquil existence was about to be disrupted in the rudest possible manner. Within a matter of months this retiring little creature unwittingly found itself spearheading an onslaught on the $120 million Tellico Dam, a campaign that succeeded in bringing notoriety to the nation's largest public utility and embarrassment to the President of the United States himself. At the heart of the issue was a legally established and immutable right to protection for endangered species. The main difficulty, as was soon to become apparent, was that this right created serious obstacles to regional economic development. The question which then arose was whether it might not be preferable to modify this immutable biotic right so as to take account of other pressing objectives, principally economic development. It is instructive to observe exactly how this dilemma was resolved.

ORIGINS

The Little Tennessee River flows northwest for 230 km from the mountains of Rabun County, Georgia, through the rugged ravines and wooded valleys of North Carolina and the rolling hills of southeast Tennessee, finally to converge with the Tennessee River at Lenoir City. Originally a free-flowing mountain stream, the Little Tennessee River has been progressively harnessed since the 1930s to provide the region with cheap hydroelectric power; when David Etnier made his discovery there were no less than 15 dams along the river and its tributaries, making it one of the US's most heavily exploited watersheds. The greatest impact on the Little Tennessee's lower reaches was made by the construction of the Fontana Dam in 1944, which led to a significant reduction in water temperature downstream. But the low temperature of the dam's tailwaters provided ideal conditions for trout, particularly along the 21 km stretch below Chilhowee Dam where clear, free-flowing waters permitted the establishment

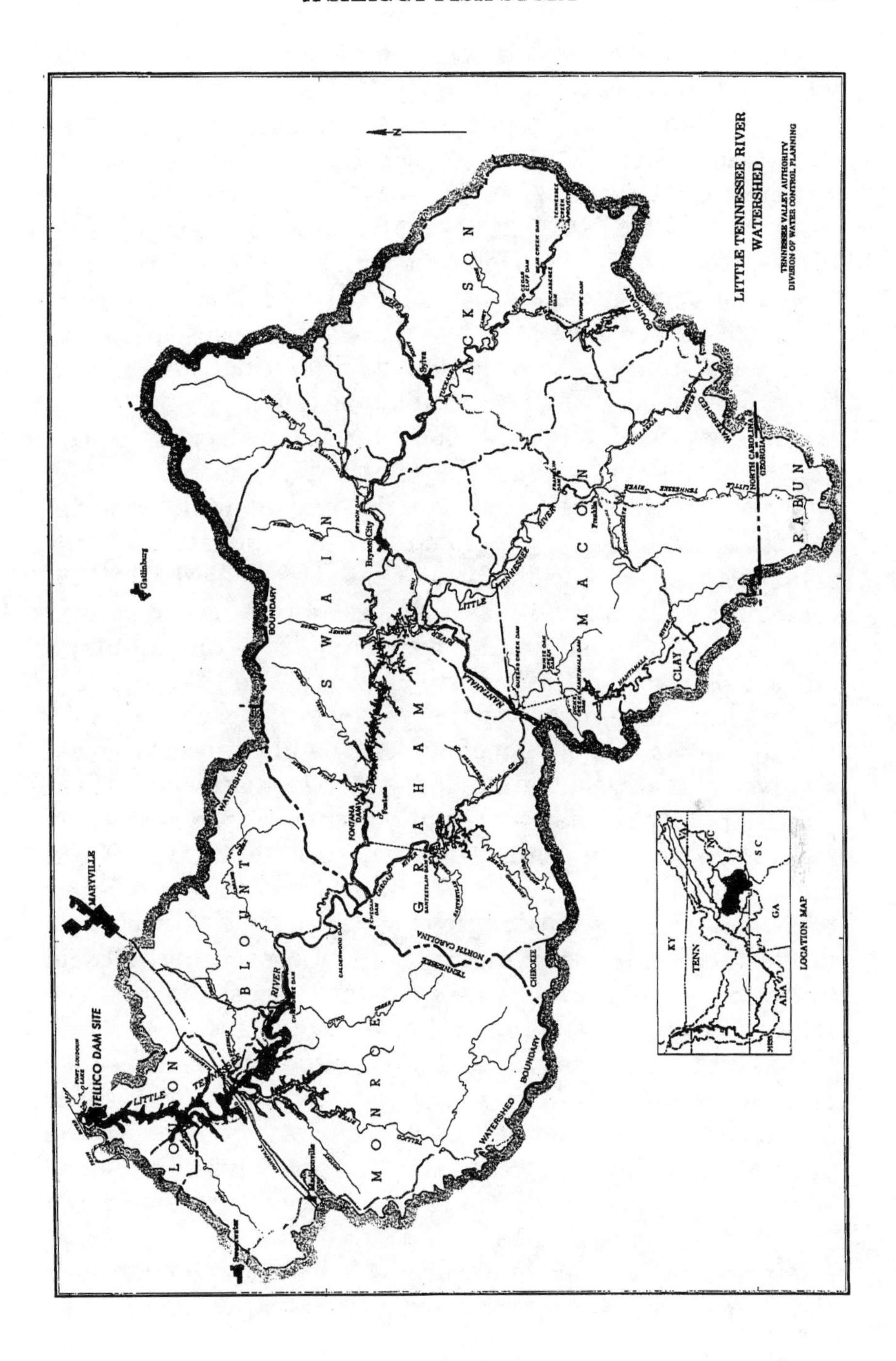
LITTLE TENNESSEE RIVER
WATERSHED
TENNESSEE VALLEY AUTHORITY
DIVISION OF WATER CONTROL PLANNING
TELLICO DAM SITE
MARYVILLE
Gatlinburg
Bryson City
FONTANA DAM
LOUDON
BLOUNT
MONROE
GRAHAM
SWAIN
JACKSON
MACON
CLAY
RABUN
CHEROKEE
LITTLE TENNESSEE RIVER
WATERSHED BOUNDARY
NORTH CAROLINA
TENNESSEE
GEORGIA
LOCATION MAP
KY
TENN
NC
SC
GA
ALA

of a recreational fishery. Indeed, the river acquired the reputation of the finest trout stream in the eastern US.

In addition to its reputation as a fishery, the Little Tennessee River valley is of considerable archaeological and historical value. A total of 14 sites have been listed or nominated for listing in the National Register of Historic Places. Contributing to this importance is the fact that the area was once inhabited by the Cherokee Indian Nation. Although most Cherokee have since come together in the Qualla Boundary reservation in the Blue Ridge Mountains, the valley remains their spiritual homeland.

The history of river development in the region is the history of the Tennessee Valley Authority. Created in 1933 by President Roosevelt to control flooding along the Tennessee River and to bring agricultural and industrial development to an exceptionally depressed region, the utility, commonly known simply as TVA, has since developed into the nation's largest public authority. It is the country's biggest producer of electricity, although two-thirds of this power now comes from coal-fired generating plants. TVA also owns over 500 000 ha of land, 800 million tonnes of coal reserves and large uranium deposits in five western states.

In 1933 all this was a dream, but the first steps towards making the dream come true were soon taken. In 1936 a master plan for the development of the Tennessee River basin was presented to Congress, setting out TVA's proposals for flood control, improving navigation, conserving soil and initiating agricultural and industrial development. The scheme was unprecedented in its scope and scale, and it was not surprising that many of the less conspicuous features went virtually unnoticed. One of these minor elements was a relatively innocuous assessment of the prospects for damming the mouth of the Little Tennessee River, which concluded that the project was not considered to be feasible at that time.

Several years later, however, TVA drew up plans to dam the Tennessee River just upstream of its confluence with the

Little Tennessee. This new proposal, the Fort Loudoun Dam, coincidentally provided the opportunity to dam the smaller river, for the newly formed reservoir on the Little Tennessee could easily be linked to the waters backed up behind the Fort Loudoun Dam by a short canal. The generating capacity of the Fort Loudoun Dam could therefore be increased, while the barrier across the Little Tennessee River could be built as a simple retaining structure without turbines. This second dam was in fact referred to as the Fort Loudoun Extension and offered the engineers a neat way of increasing the efficiency of the design. Certainly, the logic of the proposal seemed flawless; Ed Lesesne, former TVA Director of Water Resources, remembers it as a "terrific project". But in October 1942, four months after Congress had approved funds for the scheme, the War Production Board stepped in to halt further construction work because of material shortages; only the Fort Loudoun Dam was completed.

Nothing more was heard of the scheme for nearly 20 years; but the extension was not forgotten, and it eventually re-emerged in 1960 when a TVA report concluded that the investment would produce marginal net economic benefits. By 1961 the scheme had even come to head the authority's list of "future multiple use projects", although there were sharp internal disagreements as to the true value of the extension. In the absence of strong supporting arguments, the economic analysis might well have proved sufficiently suspect to deflect investment funds towards other projects. Quite by chance, however, other forces chose this very moment to come into play. January 1961 saw the inauguration of John F. Kennedy as President, and as part of his campaign to stimulate the US economy he asked all federal agencies to submit proposals for new public works projects. Keen to see early results, Kennedy demanded a speedy response and TVA duly presented him with three schemes. One was the Fort Loudoun Extension, relabelled for the occasion as the Tellico Dam.

Now that the extension had become a formal proposal,

TVA began to investigate ways of modifying the scheme so as to produce greater economic benefits. By 1963 it had become clear that the most promising idea was to embark on a major land development project based on the promotion of industrial, commercial and residential growth. To this end TVA proposed purchasing 15 400 ha of property along the Little Tennessee River, 6700 ha of which would be inundated by the reservoir. It was estimated that reselling 2000 ha of prime industrial land created by the development would bring in $10 million, the first return on the projected total investment of $41 million.

Thus it was that in contriving to ensure that a minor extension to an existing dam should produce clear economic benefits, a simple river impoundment scheme quickly evolved into a multi-million-dollar shoreline development project. To many civic groups and local officials, this sudden transformation carried the prospect of bringing economic prosperity to an area suffering from agricultural stagnation and high unemployment; it was a venture to be warmly welcomed. One of the proponents of the project, Mayor Charles Hall of Tellico Plains, saw the development as a huge step forward for the community even though, as so often, progress had its price. In this case it was to be paid by those families who would be forced to leave their homes, perhaps comforted in the knowledge that, as Hall put it, their "sacrifice was for the benefit of future generations".

But the fact that a certain degree of sacrifice was to be expected from the local community was not appreciated by a number of its citizens, least of all the likely victims, as TVA was about to discover. Although the authority's efforts had served to strengthen the economic arguments for the dam, they had also ensured that the project would be far more intrusive, both environmentally and socially, than first envisaged. Whereas objections to the original proposal had come mainly from local trout fishermen, the new, improved scheme effectively triggered opposition from a far wider range of interests and was eventually to lead TVA into the biggest controversy in its history.

RESISTANCE

In order to finance a major public works project, a utility like TVA has to be granted specific appropriations by the US Congress; each year an amount will be included in TVA's budget to cover the projected cost of carrying out the works. Competition for public funds is intense, and securing influential support in Washington is often crucial to a proposal's success. With its long history as a public utility, TVA is naturally well-versed in such matters and diligently assembled a powerful lobby behind the Tellico Dam from the earliest days of the proposal. Among the more influential members of this lobby were several leading political figures from Tennessee: Governor Frank G. Clement; Senators Albert Gore and Ross Bass; and Representatives John Duncan and Joe L. Evins. Together with Senator Lister Hill of Alabama, these men constituted a most impressive group. Evins and Hill were important members of the House of Representatives and Senate Appropriations Committees respectively, while John Duncan represented Tennessee's Second District where the bulk of the development was to be carried out and which would therefore gain the greatest financial benefit from the project.

The first hearings on the project got underway in 1965, but by this time a sizable anti-dam coalition had begun to make its presence felt. This grouping included: the 1200-member strong Association for the Preservation of the Little T; Cherokee Indians who objected to the flooding of their ancestral home; and Senator Allen Ellender, chairman of the Public Works Appropriations Committee, who disapproved of TVA's excursion into speculative development. Lyndon Johnson, who had succeeded to the presidency after Kennedy's assassination in 1964, also obstructed the proposal by insisting that TVA choose between the Tellico Dam and its proposed Tims Ford Dam on the Elk River in order to reduce public spending. Fortuitously, the Elk River project was to be situated in the district represented by Joe Evins, and faced with a choice between the two schemes he was quickly persuaded by the arguments against the Tellico Dam. This switch in allegiance was crucial, for the influential position of Evins proved decisive in ensuring that the Tims Ford Dam rather than the Tellico project gained Congressional support. TVA chairman Aubrey J. Wagner nevertheless persisted in his support for the scheme, and 1966 saw the same forces in action again, although this time with the significant difference that Evins, his local project now approved, agreed to back the Tellico Dam. The battle with the ever-expanding anti-dam coalition was fierce, but with Evins's new-found enthusiasm to the fore Congress eventually agreed to fund the project. At last construction could begin.

ANATOMY OF A PROJECT

The scheme to which Congress had given its approval consisted of four main elements: the main retaining dam and associated dykes; a canal to divert the impounded waters into Fort Loudoun Lake; alterations to the existing road and rail network; and a new community along the shores of the Tellico Reservoir. The dam itself was to be a combination mass concrete and earth embankment, 987 m long and 39 m tall. Three radial gates were to be built into the concrete section to help regulate the water level, which under normal conditions would vary over a range of 2m. Behind the dam the impounded waters would create a reservoir 53 km long stretching back to the foot of the Chilhowee Dam. For much of this distance the reservoir would be about 1 km in width, although at two points in its upper reaches it would broaden out to about 2 km.

The diversion canal linking Tellico Reservoir with Fort

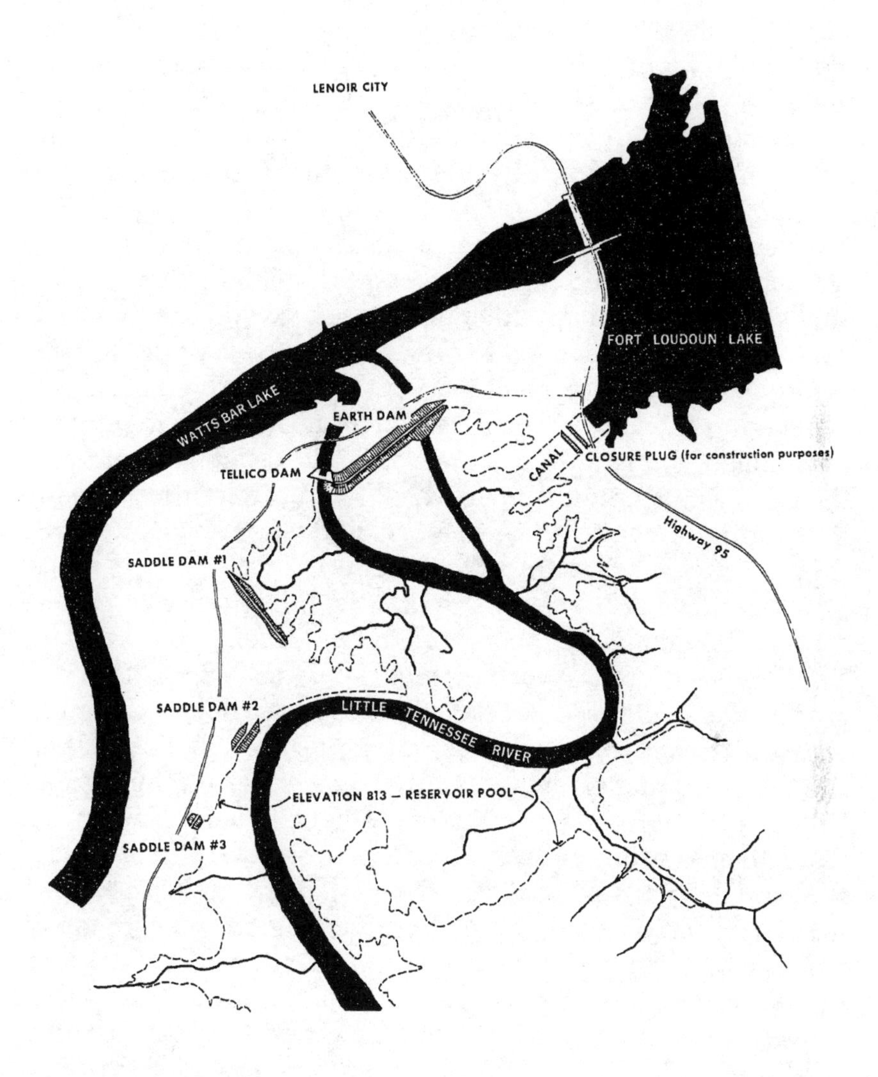

Plan of the Tellico Dam

Loudoun Lake was to be about 300 m long and 180 m wide. Because the canal would cross the path of Highway 95 the construction of a viaduct, one of 13 new bridges, would be required. Two of these would be substantial undertakings: the L & N Railroad Bridge and the adjacent Highway 411 Bridge at Vonore. In addition, 124 km of roads and 5 km of railway track would have to be constructed or relaid.

Diverting the waters of the Little Tennessee River to Fort Loudoun Lake would produce an additional output from the dam's turbines of about 200 million kWh per year, enough to heat about 20 000 homes. However, this power would not be available as a higher peak load output since the maximum generating capacity of the Fort Loudoun Dam was to remain unchanged. Instead the increase would serve to augment the dam's output at times when it was operating below its maximum capacity, thereby raising the average annual capacity factor from 45 to 62 per cent.

The major feature of the project was the creation of a new community along the shores of the reservoir. To be called Timberlake (after Henry Timberlake, the British army officer who explored the Little Tennessee Valley in the mid-eighteenth century) the development was a classic example of the "utopian" model towns favoured by planners of the period. Villages containing a number of residential neighbourhoods were to make up the community, the clusters of housing separated by trees, parks and open spaces and linked by a network of walkways. Road traffic was to be segregated from pedestrian and play areas, and strong emphasis was placed on the use of advanced telecommunication systems in traffic control, security monitoring and the operation of public services. Up to 6000 jobs were to be provided through carefully controlled industrial development – "the unified management approach to total community development". The development programme was to extend over a period of 20 years, with the ultimate goal of producing "a high quality living, working and recreational environment for about

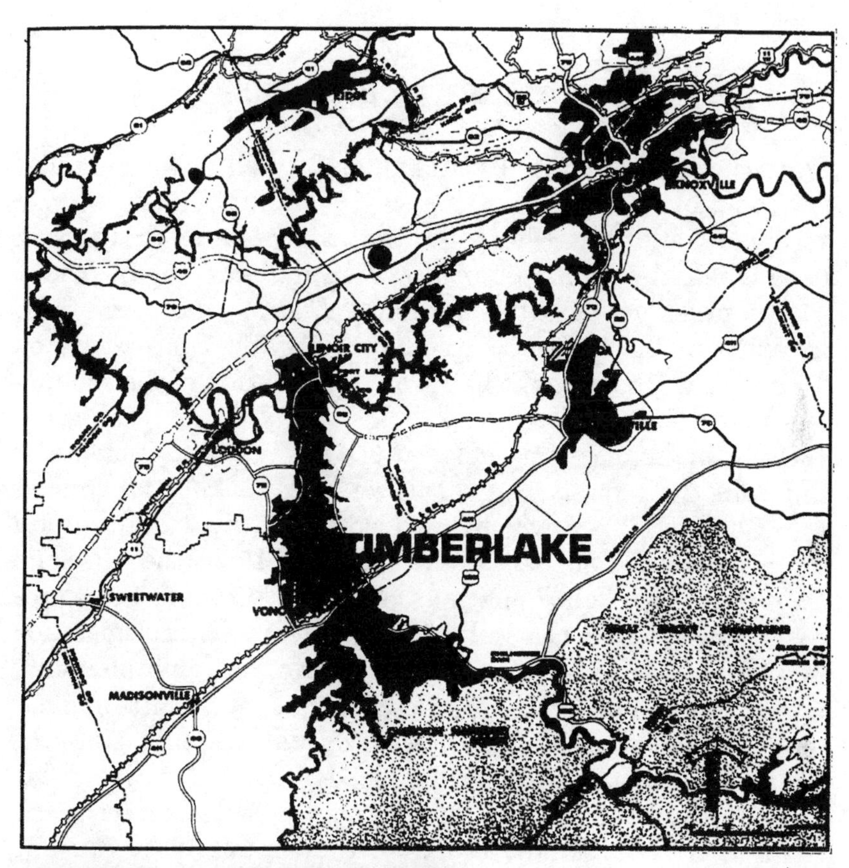

Plan of the Timberlake development

30 000 residents". Aubrey Wagner felt that Timberlake was "a tremendous idea".

AN ATTACK OF HICCUPS

Construction began in 1967. The dam itself was scheduled for completion in the spring of 1970, but within a year serious problems arose in funding the project. This was a time when the demands of the Vietnam war began to put severe pressure on all sectors of public expenditure. With inflation pushing construction costs up and appropriations for public works projects being cut, completion of the scheme was deferred to 1973.

Quite by chance, this delay was to offer the opponents of the dam the opportunity they had been looking for, and it was Congress itself which provided them with the powerful weapons needed to attack the dam. In 1969, the National Environmental Policy Act was approved. Coming into force on 1 January 1970, this milestone in environmental legislation required all federal agencies to prepare an environmental impact statement for any proposal "significantly affecting the quality of the human environment". TVA argued that, since the Tellico project was initiated before the legislation was introduced, it was exempt from the new requirements. Nevertheless, the authority attempted to defuse the opposition by drawing up an environmental impact statement. This was a relatively brief document and satisfied neither the environmentalists nor a number of government agencies, including the Tennessee Department of Conservation, the Environmental Protection Agency and the Department of the Interior. The outcome was that in August 1971 the opponents of the dam, assisted by the Environmental Defense Fund, brought an action against TVA for non-compliance with the provisions of the Act.

The case was heard in early 1972 by the US District Court for Eastern Tennessee. Finding that the National Environmental Policy Act did not provide for the exemption of projects already underway when the legislation came into effect, the court granted a preliminary injunction suspending further work on the dam pending the preparation of an appropriate environmental impact statement. TVA, contesting the judgment, duly appealed to the Sixth Circuit Court of Appeals, but the ruling of the lower court was upheld. After spending $29 million on land acquisition, completing the concrete section of the dam and building a four-lane road bridge over the non-existent reservoir, all construction work ground to a halt. It was to take TVA a further 21 months to draw up a statement detailed and comprehensive enough to satisfy the District Court, and it was not until late 1972 that the injunction was lifted and work could resume.

Things seemed to be going TVA's way at last, for not only could construction work resume but a prestigious ally had been recruited to help finance Timberlake – the Boeing Corporation. On the basis of a feasibility study carried out jointly with TVA in 1971, the company had agreed to invest $250 000 in the project. The involvement of a commercial organization with the status of Boeing had the immediate effect of conferring considerable credibility on the project, and TVA could be forgiven for viewing the future with some optimism. Yet again, events were to prove such optimism unfounded. Even as the District Court was lifting the injunction, developments were afoot which were to escalate the conflict even further.

A FISH OUT OF WATER

It is often timing which determines the political significance of events. As an organization intimate with the facts of political life, TVA understood the value of doing the right thing at the right time only too well. But this principle applied equally to its opponents. One of the experts who testified on behalf of the anti-dam coalition was Dr David Etnier, a zoologist from the Department of Zoology at the

University of Tennessee. Etnier was asked to serve as an expert witness not only because of his unequivocal opposition to the dam, but also because his main field of study was the fish faunas of large streams and rivers. The testimony which he made to the District Court emphasized the point that the Little Tennessee River system was of exceptional natural value, providing a habitat for perhaps 10 rare and endangered species. Moreover, it was likely that some of these species inhabited the stretch due to be inundated by the Tellico Reservoir.

When the case went to appeal, David Etnier was again entered on the list of expert witnesses willing to appear for the coalition, but since his first testimony had been based solely on previously published data, he decided to visit the site and survey the fish population in the hope of finding one or more of the threatened species. On 12 August 1973, just before the appeal hearing, Etnier and a colleague, Dr Robert A. Stiles, began their investigation at Coytee Spring, 11 km upstream from the dam. Remarkably, the very first fish that David Etnier saw that day turned out to be an unknown species. He managed to catch it and recognized the fish as a member of the perch genus, *Percina*. It was tan coloured, speckled, about 6 cm long and, it must be said, rather undistinguished as fish go. Subsequent taxonomic study showed the specimen to be a distinct species, although closely related to the stargazing darter, *Percina uranidea*, and the size of the population was estimated at 10–15 000. Two years later, when the fish was officially accredited with separate status, it was given the scientific name *Percina tanasi* after the nearby village which served as the capital of the Cherokee Nation until 1725. But its common name was more evocative: snail darter, reflecting the creature's favorite diet. In scientific terms it must be acknowledged that the discovery was not of major significance; in Tennessee alone some 90 species of darter were known to exist (including about 50 in the Tennessee River system alone) and new species were being described at an average rate of one a year. But there is nevertheless some irony in the fact that David Etnier was not called to testify in the appeal case. Despite the discovery, no one then realized that the Tellico Dam issue would never be the same again.

On its own, a rather nondescript fish would have stood little chance against a multi-million-dollar public works project. At that very moment, however, Congress was putting the finishing touches to a piece of legislation which effectively armed the little fish to the gills – the Endangered Species Act. Signed into law on 28 December 1973, the Act represented the response of Congress to increasing public pressure for the protection of endangered species. Previous legislation in 1966 and 1969 had required the Secretary of the Interior to draw up lists of endangered species, but these statutes provided no formal protection. This omission was subsequently recognized by Congress which had noted the decisive role played by human activities in the increasing rate of extinctions. To confer the necessary protection, the 1973 Act introduced the principle of conserving the ecosystems upon which the endangered species depended. This was to be achieved through a two-stage process.

First, the Secretary of the Interior was required to list all "endangered" and "threatened" species together with their respective "critical habitats". An endangered species was defined as any species that is in danger of extinction throughout all or a significant part of its range, whereas a threatened species was described as any species likely to become an endangered species in the foreseeable future. A species' critical habitat was not defined in the Act, but a later administrative ruling by the Secretary of the Interior interpreted the term as meaning any air, land or water area together with their constituent elements, the loss of which would appreciably decrease the likelihood of the survival and recovery of the species or a distinct proportion of its population.

Second, all federal departments and agencies were to take "such action necessary to insure that actions authorized, funded, or carried out by them do not jeopardize the continued existence of such endangered species or result in the destruction or modification of the habitat of such species" (Section 7). The wording of this part of the Act was to prove vital, for it evidently provided mandatory and absolute protection to all endangered and threatened species and their habitats from any federal and federally authorized activities.

The significance of these provisions did not go unnoticed by those still campaigning to stop the Tellico Dam. Once the distinct taxonomic status of the snail darter had been recognized (although it was not until January 1976 that this was formally established by publication of the description of the species). Joseph P. Congleton, Zygmunt J.B. Plater and Hiram G. Hill petitioned the Secretary of the Interior, Rogers C.B. Morton, to list the fish as an endangered species, and under the provisions of the Endangered Species Act he had little choice but to do so. Material accompanying the claim showed that the snail darter had only been found in a single stretch of water, despite the fact that the rivers and streams in the area had been well collected, and that a variety of interested parties consulted on the proposal, including TVA and the State of Tennessee, could not provide any countervailing evidence. Noting that "more than 1000 collections in recent years and additional earlier collections from central and east Tennessee have not revealed the presence of the snail darter outside the Little Tennessee River", the Secretary duly designated the fish as an endangered species on 9 October 1975. Six months later, and more importantly for the issue at hand, the snail darter's critical habitat was defined as lying between river mile 0.5 and river mile 17 on the Little Tennessee River – entirely within the area to be impounded by the reservoir. The implications of this were not lost on the Secretary of the Interior. In designating the fish as an endangered species the Director of the US Fish and Wildlife Service added that it occurred:

> only in the swifter portions of shoals over clean gravel substrate in cool, low-turbidity water. Food of the snail darter is almost exclusively snails which require a clean gravel substrate for their survival. The proposed impoundment of water behind the proposed Tellico Dam would result in total destruction of the snail darter's habitat.

And in designating the Little Tennessee River as its critical habitat, the acting Director of the Service spelled out the consequences of this decision, declaring that:

> all Federal agencies must take such action as is necessary to insure that actions authorized, funded, or carried out by them do not result in the destruction or modification of this critical habitat area.

The prospects for the dam were not encouraging.

As was to be expected, TVA had not remained inactive during these developments. It realized at an early stage that the threat to the dam posed by the snail darter could be nullified if other populations of the fish were to be discovered. The authority accordingly carried out a comprehensive survey of potential sites in the region, covering more than 60 watercourses at a cost reputed to be close to $1 million. But at no other location was a single snail darter found, and TVA's determined efforts to persuade the Secretary of the Interior not to designate the Little Tennessee River as its critical habitat failed. Discussions had also taken place with the Department of the Interior's Fish and Wildlife Service to see whether some compromise over the project might be possible. But TVA resisted proposals to make major modifications to the scheme, agreeing only to transplant the entire snail darter population to another site. (The chief official from the Department of the Interior at the negotiations recalls that at the first meeting TVA's lawyer sat across the table and said, "Let's get one thing straight; the gates of the dam will be closed on January 1st 1977".) In fact, between June 1975 and February 1976 TVA itself transplanted a number of snail darters to the Hiwassee River in southeastern Tennessee, the first of four such trials. But the Secretary remained unconvinced by this approach, seeing little evidence to refute the conclusion that biological and other factors precluded the establishment of a successful breeding population in the Hiwassee River (to which the fish had had access in the past).

TVA was not the only body at odds with the Department of the Interior. Even while the Secretary was considering the petition to list the snail darter as an endangered species, the House of Representatives Appropriations Committee was voting an additional $29 million for construction work on the dam during 1976, unequivocally declaring that "the project . . . should be completed as promptly as possible". The outcome of the vote undoubtedly reflected sympathy for TVA's argument that the Endangered Species Act did not prohibit the completion of a project that had been authorized, funded and substantially constructed before the Act was passed. President Ford evidently concurred with this interpretation, for he signed the Appropriations Bill into law in December 1975, two months after the snail darter had been listed as an endangered species.

By now, however, the uncertainty surrounding the project was beginning to take its toll. From its soundings in Washington, Boeing concluded that Congress was unlikely to approve funds for the Timberlake development in the near future and announced its withdrawal from the project in early 1975. TVA subsequently found the corporation's assessment to be correct, for despite intensive lobbying for the model town, Congress refused to appropriate the necessary funds, leaving the scheme shorn of its principal economic justification. At this stage many observers thought that the loss of the Timberlake development, combined with the complications introduced by the Endangered Species Act, might prompt the cancellation of the entire project. But TVA was not to be deterred. General manager Lynn Seeber reiterated the authority's determination to continue construction work, declaring that "the dam will be built".

Seeber's confidence could be explained in part by TVA's success in gaining an important new ally – Ray Blanton, the new governor of Tennessee. Notwithstanding the fact that in 1972 Blanton had called the project a "bureaucratic blunder", on becoming governor he was soon persuaded of the dam's merits by Representative Joe Evins. He declared that he now accepted TVA's claims and that there would be no opposition whatsoever from his administration. In fact the lobbying of key politicians was becoming ever more intense, and in these circumstances the new-found political support from the state administration more than made up for the loss of Boeing's financial commitment.

It was clear to the opponents of the project by early 1976 that the only hope of halting work on the dam lay in litigation. Despite the Secretary of the Interior's designation of the snail darter as an endangered species and the reservoir site as its critical habitat, Congress seemed intent on continuing its appropriations for the project and was being pressed all the way by a powerful state coalition. The time had come for the snail darter to be ushered onto centre stage.

A STAR IS BORN

The judicial battle to save the snail darter was initiated by a coalition of opponents to the Tellico Dam. The group, consisting of the Association of Southeastern Biologists, the Audubon Council of Tennessee and a number of concerned citizens, brought a case in February 1976 to halt completion of the dam and prohibit the impoundment of the reservoir on the grounds that such action would violate the Endangered Species Act by causing the extinction of the snail darter. At the hearing in the Eastern District Court of Tennessee on 29 and 30 April, the opponents of the dam contended that impoundment of the Little Tennessee River would cause the extinction of the snail darter by drastically modifying its critical habitat; it thereby constituted a violation of the Endangered Species Act. TVA's defence was that the Act did not apply to the Tellico Dam since the project had been approved before the legislation had been passed and before the discovery of the snail darter. It also stressed that the project was now 80 per cent complete and had already cost $78 million, $53 million of which was claimed to be irrecoverable. Judge Robert L. Taylor, after considering the arguments for more than three weeks, entered his judgment on 25 May. Although agreeing with the plaintiffs that the impoundment of the river would adversely modify, if not completely destroy, the snail darter's habitat, he accepted TVA's arguments and refused to issue an injunction to prohibit further work on the dam:

> At some point in time a federal project becomes so near completion and so incapable of modification that a court of equity should not apply a statute enacted long after inception to produce an unreasonable result .... Where there has been an irreversible and irretrievable commitment of resources by Congress to a project over a span of almost a decade, the Court should proceed with a great deal of circumspection. . . . If plaintiff's argument were taken to its logical extreme, the Act would require a court to halt impoundment of water behind a fully completed dam if an endangered species were discovered in the river on the day before such impoundment was scheduled to take place. We cannot conceive Congress intended such a result.

Given the green light to proceed, TVA needed little further urging: it promptly introduced a three-shift, round-the-clock work schedule in an effort to complete construction as soon as possible. As part of this operation, all vegetation below the 250 m contour was cut and then buried in trenches or burned. Opponents of the project saw this move as an attempt to present an appeals court with a *fait accompli* and as final proof of TVA's obduracy. Worse news was to follow. Less than a month after the District Court's judgment, the Senate and House Appropriations Committees voted to continue federal funding of the project. Approving the full budget request of $9 million, the Senate committee declared that it did not regard the Endangered Species Act as an obstacle to impoundment and recommended that "this project be completed as promptly as possible in the public interest".

To the dam's opponents the public interest could best be represented by halting construction work immediately. It therefore came as no surprise when they appealed against the District Court's ruling, bringing a case in the Sixth Circuit Court of Appeals. The familiar arguments were again expounded, with the plaintiffs further contending that the District Court had abused its discretion by refusing to issue an injunction in the face of "a blatant statutory violation". This time their persistence was rewarded. In its judgment of 31 January 1977, the Court of Appeals began by accepting the District Court's opinion that impoundment would almost certainly cause the extinction of the snail darter. It then departed from the lower court's reasoning, finding that the Endangered Species Act made no allowance for the degree of a project's completion: abandoning the dam was therefore appropriate if it secured the continued existence of the snail darter.

Two key issues were considered in the ruling: whether the Endangered Species Act was to apply to projects already underway, and whether Congress's continuing appropriations represented an implied exemption from the Act. To the Court of Appeals, TVA had clearly failed to take "such action necessary" to ensure that its "actions" did not jeopardize the existence of the snail darter, as required by Section 7 of the Act. Reviewing the legislative history of the Endangered Species Act, the court found no indications to support TVA's argument that the word "actions" did not apply to the terminal phases of projects already underway when the Act became law. The actions currently being taken by the authority were, on its own admission, likely to destroy the snail darter population. That Congress had repeatedly approved funding for the Tellico Dam made no difference; this amounted only to "advisory opinion" concerning the proper application of an existing statute.

In the light of these findings the Court of Appeals concluded that the lower court had erred by not issuing an injunction, despite the fact that this would have led to the loss of millions of dollars of public funds. It therefore instructed the District Court to issue a permanent injunction prohibiting further work on the Tellico Dam, directing that the injunction could only be lifted if Congress passed legislation to exempt the project from the provisions of the Endangered Species Act or if the snail darter was removed from the list of endangered species or if its critical habitat were to be redefined.

The decision of the Court of Appeals was not only hailed as a great victory by those fighting to protect the Little Tennessee Valley from impoundment, it succeeded in elevating the conflict to a *cause célèbre*, for by now the case had come to be seen as the "little fish versus the big dam". Saving the helpless snail darter from the rapacious public utility dramatically symbolized the greater issue of protecting nature from remorseless human exploitation. The image of an obscure fish holding back the operation of an almost completed dam was just too much for the popular media to resist.

It was also too much for the proponents of the dam to bear. Without delay TVA appealed to the Supreme Court against the ruling. It also reopened discussions with the Department of the Interior in an attempt to persuade the Fish and Wildlife Service to delist the Little Tennessee River as the snail darter's critical habitat, and at the same time launched a new political offensive aimed at securing explicit legislative approval for the scheme. Following an initiative by Governor Ray Blanton, Representative John Duncan proposed that the Tellico Dam be explicitly exempted from the provisions of the Endangered Species Act. The energy crisis brought about by the Arab oil embargo conveniently provided additional support for this proposal; and when he appeared before the Senate Subcommittee on Natural Resource Protection in July

1977, TVA's Lynn Seeber was not slow to point out that the diversionary flow to the Fort Loudoun Dam would generate much-needed hydroelectricity. Seeber's enthusiasm for the project was underlined by his claim that the additional power would be worth $3.5 million a year – which was considerably more than earlier projections made by the authority. More importantly, it was also considerably more than the estimate made by the US General Accounting Office in its report on the dam's benefits which was published that summer. In fact the report concluded that other alternatives were feasible and that further analysis was needed.

This difference of opinion between TVA and the General Accounting Office was to form the prelude to two rather more acrimonious conflicts which neatly illustrated the inexhaustible capacity of the issue to generate discord. As a federal agency, TVA was able to secure the services of the Department of Justice to plead its case in the Supreme Court. Attorney General Griffin Bell argued that, since the Department of Justice must be seen to represent government opinion, the administration should support TVA's efforts to build the Tellico Dam. The Secretary of the Interior, Cecil Andrus, took a rather different view of where government loyalty should lie. His concern was for the Endangered Species Act, and he was adamant that the statute precluded the completion of the dam. This domestic difficulty within the administration was finally resolved by a device which was unusual, to say the least. It was agreed that the Department of Justice should prepare the brief setting out TVA's case (which it duly filed with the Supreme Court on 25 January 1978) but that attached to the brief should be an appendix, drafted by the Department of the Interior, taking the opposite position and arguing against the dam. Such a conspicuous division within the government inevitably led to the involvement of the President himself. Jimmy Carter had been in office for a year and was keen to give his administration an environmental bias. He pressed Bell to switch positions and oppose the completion of the dam. However, the Attorney General refused to do so, and the infighting, apparently some of the most bitter of Carter's entire Presidential term, continued.

By now, the pressures were such that even TVA was beginning to divide over the issue. The first cracks in the edifice appeared in April 1978 when a new board member, S. David Freeman, suggested to Andrus that a dry dam for emergency flood control together with industrial development without impoundment should be considered as an alternative. This was clearly a radical proposal for a TVA representative, and it was certainly not a move calculated to endear Freeman to Chairman Wagner. But the appointment of Freeman to the three-member board by President Carter in August 1977 was inevitably going to mark a new phase in the course of the authority. As an energy adviser to two presidents, Freeman's unconventional views were well known: he advocated raising energy prices to incorporate environmental damage costs, expressed concern about the risks of nuclear power and was a frequent critic of big business, particularly the major oil companies. Boardroom clashes were inevitable. They were to be short-lived disputes, however, for after Wagner's retirement in 1978, Freeman became chairman. By then the third board member, William Jenkins, had resigned out of frustration, so within the space of a year Carter had succeeded in completely changing the make-up and complexion of TVA's top management. Moreover, the Fish and Wildlife Service had stated in no uncertain terms that it was not prepared to consider delisting the snail darter or its critical habitat until the question had been resolved in Congress or the courts. In these circumstances both Carter and the dam's opponents must have thought that the project was as good as dead.

That, however, remained to be decided by the Supreme Court. And it was fitting that one of Aubrey Wagner's last duties as Board Chairman was to bring TVA to the highest US court to fight the Sixth Circuit's injunction against completing the dam. The case opened on 18 April

1978 with the Attorney General, acting for TVA and doing his best to ignore the appendix to his brief, arguing for the dam's completion: construction was virtually finished, the project had been initiated before the introduction of the Endangered Species Act and the survival of the snail darters transplanted to the Hiwassee River effectively ensured that the fish was no longer at risk. The case for the defence similarly trod familiar ground: the Endangered Species Act allowed no exemptions for partially completed projects and the dam would indisputedly modify the critical habitat of a species registered as endangered. Not only that, but the prosecution's brief, as defence counsel maliciously pointed out, also argued against the project.

It took the nine justices a month to consider their verdict and on 15 June 1978 the final judicial pronouncement on the Tellico Dam was made: by a majority of 6–3 the court ruled that impoundment of the Little Tennessee River would indeed constitute a violation of the Endangered Species Act. In reaching its judgment, the court first studied the history of endangered species protection legislation in order to determine the exact intent of Congress. From this the majority opinion concluded that the "plain intent of Congress in enacting the Endangered Species Act was to halt and reverse the trend towards species extinction, whatever the cost". But did the continued appropriations for the dam by Congress amount to an implied repeal of the legislation? No, for this would violate the "cardinal rule . . . that repeals by implication are not favored". And since there was no dispute that impoundment of the Little Tennessee River would destroy the designated critical habitat of the snail darter, there was an "irreconcilable conflict between operation of the Tellico Dam and the explicit provisions of §7 of the Endangered Species Act". The only reasonable remedy was a permanent injunction against completion of the project, and the Supreme Court accordingly affirmed the earlier injunction of the Court of Appeals.

THE BEGINNING OF THE END OR THE END OF THE BEGINNING?

It was a famous victory and the opponents of the project were justifiably elated. The Tellico Dam was sentenced to become a white elephant, bearing witness to TVA's environmental insensitivity. For the utility itself, some alternative course of action had to be devised. TVA set about this task, in cooperation with the Department of the Interior, by developing and analysing four alternative strategies: proceeding with the dam as originally intended (on the assumption that some feasible solution to the snail darter problem could be found within the constraints of the Endangered Species Act); building a dam at an alternative location (the Tellico River); developing the project area without creating a permanent reservoir; and selling the land at the highest possible price. A report evaluating these alternatives was published in August 1978 and, although it carefully omitted recommending a preferred course of action, a number of interesting conclusions emerged.

First, overall benefits would exceed the remaining costs of completion for both the reservoir and the river development alternatives, though neither would provide a positive return if being built from scratch in 1978 using the newer evaluation methodology and the prevailing interest rates. Second, it transpired that new safety standards recommended by the Bureau of Reclamation (particularly the ability of structures to withstand the so-called "superflood") would probably require substantial additional expenditure if the dam was completed as intended, so that instead of the projected figure of $19.4 million to cover the outstanding construction work, $33.9 million would be necessary, giving a final total cost of $143 million. Third, the income likely to be received by selling off the land to the highest bidder would exceed the estimated benefits of the Tellico Dam: the value of the land if used for agriculture would be $20–45 million, and investment costs of up to $50 million would be saved. In this connection it is important to appreciate that the $109 million expended on the project up to that point would not simply be wasted as was so often claimed during the conflict. About 80 per cent of the development costs had been incurred in land acquisition and road construction. The land could be resold at a handsome profit and many of the new roads and bridges were valuable additions to the region's infrastructure. Only about $22 million had been absorbed in the construction of the dam itself. Finally, at the press conference held to introduce the report, an astute journalist raised an interesting point:

> Q: Mr Secretary, if any decision is postponed long enough, taking into account the decrease you are noticing already of snail darters in the Little Tennessee, isn't it entirely possible that they will disappear altogether and then you can just build the project as originally intended?
>
> Robert L. Herbst (Assistant Secretary, Department of the Interior): I suppose that is a possibility. That happened to the Tecopa Pup Fish in California. Somebody built a bath house on top of that mineral spring and wiped it out, and it became the first species that was delisted because it became extinct.

The instigation for this question was a footnote in the TVA report on the status of the original population of snail darters in the Little Tennessee River and that of the transplanted fish in the Hiwassee River. Surveys indicated that the numbers of snail darters in the Little Tennessee River had drastically declined, from 10–15 000 when the fish was discovered in 1973 to just 237. This was primarily due to the effect of the earth dam which had blocked the passage of snail darters from Watts Bar Lake to their spawning grounds in the Little Tennessee River. Removal of this dam was therefore essential if the original population was to survive. In the Hiwassee River, by contrast, surveys indicated that the original population of 710 had increased to about 2000 in the space of three years. For the dam's supporters this was a positive development. But things were not quite that simple. As the report explained, pesticide run–off from agricultural land was a potential threat to the transplanted population, and copper mining in the area required the shipment of sulphuric acid by rail along the Hiwassee River gorge. Eight derailments had occurred since 1971 and a major acid spill could decimate the snail darter population.

That was not quite the end of the matter. Indeed, it soon became apparent that the real action was only just beginning, for if the snail darter's critical habitat was no longer threatened with substantial modification, the Endangered Species Act most certainly was. In fact to many of the dam's proponents, the prospect of a $119 million project lying redundant was reason enough to seek the repeal of the entire statute as a major impediment to economic development. Far from being over, hostilities simply shifted to a different battleground.

CHANGING THE RULES

Once it became clear that the political lobby behind the Tellico Dam had no intention of acknowledging defeat, the only question that remained was the exact form that the attack on the Endangered Species Act would take. The deal that was eventually worked out after intense bargaining behind closed doors did indeed represent a major modification to one of the Act's original principles, but the proposals and the reasoning behind them were rather more subtle than they at first appeared. Senators John Culver of Iowa (Democrat) and Howard Baker of Tennessee (Republican and Minority Leader in the Senate) agreed to sponsor a bill amending certain provisions of the Endangered Species Act. The thrust of the bill was to establish a cabinet-level body, the Endangered Species Committee, to rule on those cases where "irresolvable" conflicts arose in federal projects which fell within the scope of the Act. Moreover, it was proposed that economic as well as ecological criteria should be taken into account in deciding whether to designate a site as the critical habitat of an endangered species. Given the contentious nature of the Supreme Court decision, many Congressmen, including several sympathetic to the conservation cause, saw the bill as a way of defusing attempts to repeal the entire Act and accepted the contingency procedure as a necessary evil. This broad front was, in fact, discernible in the bill's co-sponsors themselves: Howard Baker regarded the amendment as a device to secure completion of the Tellico Dam in his home state, while John Culver was an ardent conservationist determined to save the essential principles of the Endangered Species Act. Because of this broad consensus Congress readily supported the bill, and in November 1978 President Carter signed the Endangered Species Act Amendments into law. A new arbiter of the snail darter's right to survive had come into existence.

The Endangered Species Committee, affectionately known as the "God Committee", was made up of five voting members (Agriculture Secretary Bob Bergland, Army Secretary Clifford L. Alexander, Chairman of the Council of Economic Advisors Charles L. Schultze, Environmental Protection Agency Administrator Douglas M. Costle and National Oceanic and Atmospheric Administration Administrator Richard A. Frank) plus presidential appointees from the affected states with a single collective vote (in the case of the Tellico Dam just William R. Willis, Jr, for Tennessee) and the Secretary of the Interior, Cecil Andrus acting as chairman. One of the two issues to be considered by the committee at its very first session held on 23 January 1979 was whether the Tellico Dam should be exempted from Section 7 of the Endangered Species Act. Under the amended legislation the committee was only empowered to exempt a project if, first, there were no reasonable and prudent alternative courses of action and, second, the benefits of the project clearly outweighed the benefits of any alternatives which would conserve the species and its critical habitat and which were also in the public interest. Needless to say, these criteria had been very carefully considered by the dam's proponents when the amendments were drafted and they were convinced that an exemption would be forthcoming – particularly since the committee represented a broad range of interests.

In its deliberations, the Endangered Species Committee reviewed the history of the project and concentrated in particular on two of the alternative options which had been evaluated in the joint TVA/Department of the Interior report: reservoir development as originally proposed and river development without the dam. The decision, taken after just 25 minutes of discussion, was unanimous. It was also remarkable, for the committee voted against granting an exemption on the grounds that, apart from having an irreversible ecological impact, the dam could not be justified in economic terms. Charles Schultze was particularly critical: there were manifestly reasonable and prudent alternatives to the dam and the costs of the project clearly outweighed its benefits. In fact, he concluded that even the costs of completing construction would exceed the total benefits of the entire project:

> The interesting phenomenon is that here is a project that is 95 per cent complete, and if one takes just the cost of finishing it against the benefits and does it properly, it doesn't pay, which says something about the original design.

Yet again the dam's opponents had cause to celebrate: both the ecological and the economic cases had been unequivocally won, judicial procedures had been exhausted and no Congressional support would be likely to be found for further amendments to the Endangered Species Act. Incredibly, the proponents of the dam still refused to accept defeat and once more began exploring new avenues for gaining the necessary authorization – proponents, it should be noted, other than TVA, whose new three-member board were determined to appear scrupulously neutral on the issue, declaring only that they would submit to the wishes of Congress. More outspoken were the Congressional representatives from Tennessee, who lost no time in giving vent to their anger, as did many of the citizens attending a public meeting in support of the project which drew a thousand people. Now, instead of openly pursuing judicial and legislative action, it was decided to resort to some astute use of Congressional procedures and intensive behind-the-scenes politicking.

BENDING THE RULES

On 18 June 1979 John Duncan, the Republican representative for Tennessee, unobtrusively attached an amendment to the $10.8 billion Energy and Water Development Appropriations Bill 1980, the annual measure authorizing the funding of a large number of federal agencies. The amendment read as follows:

> *Provided*, That notwithstanding the provisions of 16 U.S.C., chapter 35 or any other law, the Corporation is authorized and directed to complete construction, operate and maintain the Tellico Dam and Reservoir project for navigation, flood control, electric power generation and other purposes, including the maintenance of a normal summer reservoir pool of 813 feet above sea level.

Duncan's purpose was clear, if not widely broadcast. In the context of an appropriations bill of such vast scope, an amendment authorizing the completion of the Tellico Dam was of marginal significance. More than that, his choice of an appropriations bill as the legislative vehicle for the amendment reflected the fact that such measures are invariably masterpieces of political compromise – they include something for everyone. Proposals taken up in this particular bill, for example, included a new water resources council supported by President Carter, a new office building for the Senate and a 5.5 per cent pay increase for Congressional staff. The Tellico Dam nevertheless remained a highly contentious issue, and with this in mind Duncan resorted to an old House of Representatives procedural ploy: his proposal was introduced with the help of the bill's floor manager only as a number, with the result that the text was not published in advance. With no forewarning as to the content of the amendment, the House was virtually empty when the measure came up for debate on 18 June. The text was not even read to the House, although it was printed in the *Congressional Record* and appeared as if it had been. With so little awareness of what was at stake, the amendment was agreed to with no discussion and without a vote. The only dispute was to be found amongst the press – as to whether the entire procedure lasted a total of 40 or 42 seconds.

It might seem curious that an appropriations bill, designed to authorize federal expenditure, could be used to introduce legislation approving the completion of a public works project. This tactic is indeed a controversial point, for House Rule XXI(2) specifically forbids the use of appropriations bills to change existing laws:

> No appropriation shall be reported in any general appropriation bill, or be in order as an amendment thereto, for any expenditure not previously authorized by law, unless in continuation of appropriations for such public works as are already in progress. Nor shall any provision in any such bill or amendment thereto changing existing law be in order.

However, Duncan was not only astute in his use of House procedures, he was careful not to request further funds for the Tellico Dam in the amendment, thereby defusing the issue of whether a further appropriation would be legally permissible. Nevertheless, the amendment did change existing laws, for not only was the general protection provided to endangered species by the Endangered Species Act specifically removed from the snail darter population in the vicinity of the Tellico Dam, but all other federal, state and local legislation as it pertained to the dam was similarly nullified. This included statutes on floodplain management, water pollution, historical monuments and structural safety. In fact the majority judgment of the Supreme Court encompassed precisely this issue: "[appropriations bills] have the limited and specific purpose of providing funds for authorized programs."

At the time, few Congressmen were aware of what had been approved, but this was to change with the consideration of the bill by the Senate. Following publication of the amendment in the *Congressional Record*, the cat was out of the bag and the debate in the upper chamber was conducted in full knowledge of what was at stake. Confidence was nevertheless high in the Tellico camp, and not without reason, for Howard Baker, Republican leader in the Senate and avowed supporter of the dam, was actively assembling support for the amendment. Moreover, the Senate has no power to amend bills passed by the House of Representatives, each item of legislation has to be approved or rejected in its entirety. The dam's supporters reasoned that an appropriations bill of such vast scope would hardly be blocked solely on the grounds of Duncan's amendment. That, however, is exactly what did happen. On 18 July, in a narrow vote, the Senate rejected the entire bill, the major bone of contention being the Tellico Dam.

But such a stalemate still has to be resolved. Under Congressional rules this is done by putting the disputed sections of the bill before a "conference", an *ad hoc* committee comprising representatives of both the House and the Senate who negotiate revised proposals likely to gain majority support, and then resubmitting the bill to both chambers. Such a procedure is inevitably characterized by intense political infighting and "logrolling" – the trading of favours for votes amongst the beneficiaries of various schemes. But even here it proved impossible to agree on a compromise proposal for the dam, and it was therefore decided to consider the amendment separately from the rest of the conference report in the House of Representatives.

The proposal was put before the House on 1 August – during the evening session of the last day of business before the summer recess. Representatives again repeated the familiar arguments for and against the dam, with the only new contribution being a discussion on the propriety of Duncan's tactics: opinions differed as to whether it was "a question of the integrity of the process of this House" (Breaux) and "a bad legislative precedent" (Conte) or whether it was "common procedure . . . under the normal rules and customs of the House and we merely responded to the invitation by the Supreme Court that Congress should act in this fashion if it wanted the reservoir to be completed" (Myers). That was, perhaps, a somewhat loose interpretation of the Supreme Court's judgment. But what the Court did infer was that the Tellico Dam would need the explicit authorization of Congress if the provisions of the Endangered Species Act were not to apply to the snail darter, and this explicit question was now, at long last, before the House for consideration. The vote came at 21.00, and in what can only be described as a return to traditional political values, the House belatedly gave its assent to the project.

As significant as the outcome of the vote was the margin of victory: 258 votes to 156. Although this might seem a remarkable result given the House's unequivocal support for the Endangered Species Act and the subsequent judicial and executive decisions, not to mention the intensive campaigning against the dam, in retrospect it seems that the House of Representatives was always sympathetic to the project. In particular, throughout the six-year battle on the proper application of the Endangered Species Act, appropriations to enable construction work to continue were approved by the House each year, including a sum of $1.8 million in 1978 for the completion and closure of the dam. This support undoubtedly stemmed from the success of the dam lobby in advancing four persuasive, if largely specious, arguments. First, much of the $120 million invested in the project would be lost if the dam was abandoned. Second, the additional electricity generated by the Fort Loudoun Dam, valued at $2.7 million, just 0.2 per cent of TVA's gross receipts from power generation, was strategically important at a time when the Arab oil embargoes had highlighted the vulnerability of oil imports. Third, the transplants of snail darters to other rivers held the promise that the species would not in fact become extinct. And fourth, the fish was not of sufficient value that its protection should be secured at any cost. As John Duncan succinctly put it in the House debate: "Should a worthless, unsightly, minute, unedible [sic] minnow outweigh a possible injustice to human beings?"

Whether the Senate would agree with this line of reasoning was a different matter. With only a narrow vote against the project after the first debate, lobbying from both sides was intense throughout the summer recess, the greatest efforts being focused on those Senators who were regarded as most likely to change their vote. The debate, held on 10 September, again showed that there were two evenly balanced and intransigent camps. On behalf of the dam's supporters J. Bennett Johnston insisted that, "If you want an energy and resources bill, you're going to have to build the Tellico Dam". But the decisive role was played by Minority Leader Howard Baker. His personal lobbying amongst fellow Republicans succeeded in persuading five of his colleagues (Senators Danforth, Dole, Domenici, Gravel and Wallop) to switch sides and support the amendment. It was enough. The final vote turned out to be 48–44 in favour of the bill.

Only one person could stop the bill now – the President himself. But the dam's supporters were not unduly worried by this possibility. As Bennett Johnson remarked: "It was a good, clean, hard fight and I think he'll accept it gracefully". But others felt differently. Cecil Andrus for one made no secret of the fact that he was pressing Carter to use his power of veto, arguing that Congress would be unlikely to muster the necessary two-thirds majority to overturn it. Influential newspapers such as *The New York Times* and *The Washington Post* were also critical of the Tellico Dam amendment. Moreover, Carter's sceptical opinion of the project was well known and the record showed that he was not afraid of using his presidential power: the previous year he had vetoed the corresponding appropriations bill because of economic and environmental objections to a number of water projects.

But Carter was also faced with two major complications. Howard Baker was known to have White House aspirations and was therefore a potential rival for the presidency in the 1980 elections. If Carter vetoed the bill he would in effect be picking a fight with Baker and would have to be sure of winning Congressional support. He could expect to be supported by the large Democratic majority in the House of Representatives, but the more evenly balanced Senate suggested a less certain outcome. A veto could also have serious consequences for the Endangered Species Act itself, as Baker had already threatened to mount a Congressional campaign against the Act when it came up for reauthorization. "If they veto the bill using Tellico as an excuse," said James Range, an aide of Baker's, "they're going to tear the Endangered Species Act limb from limb over in the House".

For Carter, such considerations finally proved decisive. On 25 September 1979, the President accepted, "with regret, this action as expressing the will of the Congress" and signed the bill into law. A White House official explained that soundings in Congress had shown that the fight "simply wasn't worth it", and revealed that "key House members" had promised not to oppose reauthorization of the Endangered Species Act and also to support funding for the water resources council if Carter put his signature to the appropriations bill. Although Carter argued that his decision was necessary in the interests of broader environmental ends, the perception that the final nail in the snail darter's coffin had been hammered in by a president who had campaigned as an environmentalist caused anger in the conservation lobby. Lewis Regenstein, head of the Fund for Animals:

> Carter was elected as a strong environmentalist. Now he's the first to concur in the first planned extinction of a species.

Brent Blackwelder of the Environmental Policy Center:

> He backed off, although he clearly had the votes to win. He had the chance to strike a devastating blow against the pork barrel and he blew it.

Ted Snyder, president of the Sierra Club:

> It is difficult to imagine a decision more calculated to alienate the environmental community, who have been your strongest supporters until now.

AFTERMATH

Carter's decision to sign the Energy and Water Development Appropriations Bill finally gave TVA the authority to complete the Tellico Dam. In anticipating this situation while the Supreme Court case was pending, Aubrey Wagner had remarked that, if the court lifted the injunction on damming the reservoir, "We'd fill it and we'd fill it damn quick". The new chairman, David Freeman, despite his unconventional reputation, obviously concurred with Wagner's view, for TVA resumed work on the dam within 12 hours of the bill becoming law. But even at this late stage the opponents of the project would not accept defeat. Environmentalists brought an abortive suit in the District Court claiming that the original injunction was legally still in effect, and then the Cherokee Indians claimed that the Duncan amendment was unconstitutional as the flooding of their spiritual homeland would deny them religious rights which were guaranteed by the First Amendment. The District Court dismissed the case on the grounds that ownership of the property in question was a prior condition for the claim, and the last of the Cherokee had been driven out of their homeland by white settlers in 1838. The Sixth Circuit Court of Appeals upheld the decision, and the Indians' final hope faded when the Supreme Court refused to hear the case.

By then, however, it was already too late. On 29 November 1979, tired of the interminable litigation, TVA closed the flood gates of the Tellico Dam and the lower reaches of the Little Tennessee River were finally beyond reprieve. With the reservoir backing up behind the dam, the remorseless process of sedimentation began which would inexorably destroy the critical habitat of the snail darter. A final rescue operation was mounted by a team of biologists to capture as many as possible of the remaining fish and transplant them to a new home. Several hundred were saved in this way, though for how long nobody could say, for the Fish and Wildlife Service estimated that it would take up to 15 years before a proper judgment could be made as to whether the transplanted snail darters were likely to prosper.

It later transpired that, joining the few remaining snail darters under the Tellico Reservoir was an important chunk of North America's cultural heritage. The extent of the loss came to light with the release of a report prepared by the Heritage Conservation and Recreation Service of the Department of Interior's Interagency Archeological Services. This had been drawn up under an agreement signed on 22 January 1979 by the agency and TVA which belatedly recognized their obligation under the National Historic Preservation Act and the procedures of the Advisory Council on Historic Preservation to document the potential impacts of the project on the cultural resources of the locality and indicate how these impacts might be avoided or mitigated. It is in keeping with the character of the issue that a draft of the report, dated 24 May 1979, was not released until after President Carter had finally put his signature to the appropriations bill. This delay almost certainly saved TVA further embarrassment, for the findings of the study were interesting, to say the least. The report concluded that the area to be impounded was "exceptionally rich" in prehistoric remains, containing over 282 archaeological sites:

> The physical records of American prehistory present in Tellico cannot be matched in any other area this size in the continent. . . . In summary archaeological resources of the Tellico project represents one of the most important resources for the study of American cultural development in the nation.

TVA maintained that once the decision to proceed with the dam was taken by Congress, no further work could be done on the draft report because it had to concentrate its resources on fieldwork in order to gain as much information as possible from the known sites. Only at a meeting with the Eastern Band of the Cherokees on 2 October 1979 did TVA agree to release the report, and this was done three days later.

Roughly the same fate was suffered by TVA's grandiose plans for developing the shores of the Tellico Reservoir. The Timberlake Community project was abandoned and industry showed little enthusiasm for a move to the area. Faced with such an embarrassing lack of interest, TVA finally decided to cede responsibility for developing the 4450 ha of shoreline designated for industrial and residential use to a specially created state agency.

POSTSCRIPT

Throughout these events, David Etnier had been quietly continuing his research on the fish faunas of large streams and rivers. He was fortunate that some of the extensive survey work could be carried out during the field trips he organized for his biology students. On the return journey of one of these trips, in the afternoon of 1 November 1980, his group stopped off at South Chickamauga Creek, a tributary of the Tennessee River some 200 km downstream from the Tellico Dam. The stream was seriously polluted by industrial discharges however, and Etnier was not at all confident of finding any interesting fish. The group nevertheless decided to try its luck and started work with a net. The first seine hauls produced enough interesting species to encourage further efforts – which was just as well, for a little later a snail darter appeared in the net. Some feverish work produced five more. What TVA had spent close to $1 million searching for had been stumbled upon by accident: a second snail darter population.

If two populations existed, might there not be others waiting to be discovered? After evaluating the characteristics of the site, a shortlist of other Tennessee River tributaries was drawn up for further investigation, and these searches did indeed produce several more specimens. The surveys indicated that native snail darter populations existed in the lower reaches of three Tennessee River tributaries in addition to the transplanted population in the Hiwassee River. One or two specimens were also taken in two other tributaries. Although most of these populations were small and therefore vulnerable, the snail darter was clearly no longer in immediate danger of extinction. On the other hand its future was hardly assured, and with this in mind a snail darter recovery plan was drawn up by a team from the Tennessee Wildlife Resources Agency, the Fish and Wildlife Service, the University of Tennessee and TVA.

It will undoubtedly be many years before the long-term viability of the species is secure, but the new discoveries were sufficiently encouraging for the Department of the Interior to be persuaded, on 6 August 1984, to remove the snail darter from its list of endangered species and reclassify it as “threatened”. Legally the move made little difference to the level of protection enjoyed by the fish, as the key provisions of the Endangered Species Act applied to both threatened and endangered species. But it is distinctly unlikely that this was of any comfort to the snail darter.

After only a brief encounter with *Homo sapiens*, *Percina tanasi* was doubtless a sadder but wiser creature.

CHAPTER THREE

RHINE BRINE

How to dispose of seven million tonnes of salt

The concept of pollution is not always correctly understood. Pollution connotes damage to the environment. That is to say, a pollutant has to come into contact with a target and be present in sufficient quantities to cause damage. If a potential pollutant does not reach an environmental target or does so only in an excessively diluted form, it will cause neither damage nor pollution.

This reasoning implies that the emission of potential pollutants into the environment need not in itself constitute pollution; it also suggests that it may not be necessary to reduce the discharge of a pollutant to zero in order to eliminate the pollution it causes. It does not mean that there is necessarily a threshold level for all potential pollutants below which no damage will be caused, since in some cases the assimilative capacity of the environment may be zero.

In practice, there are many reasons why the ideal goal of zero pollution will not be regarded as a feasible option. Whether for economic, technical or other reasons, many activities generate emissions which exceed the capacity of the environment to assimilate the pollutants. The question of what action should be taken then arises. In most cases abatement techniques will usually be available to reduce the emissions to some extent. The problem is that the more emissions are reduced, the more expensive further reductions become; relatively simple measures such as mechanical filters might, for example, reduce untreated emissions by 50 per cent, but highly sophisticated treatment equipment might be required to secure additional reductions. In economic terms, the point at which further reductions become unwarranted is reached when the costs of reducing the emissions increase to the point at which they equal the value of the environmental damage which would be prevented by the abatement measures. In other words, the optimum level of pollution occurs when the costs of further control measures would exceed the value of the environmental benefits gained. To be precise, the optimum level of pollution will be reached when the marginal control costs increase to a level where they equal the marginal damage costs. At this point the net economic benefits from the activity will be maximized.

The theory is impressive. It does not take long, however, to realize that it raises all sorts of awkward problems. In the first place, economics takes a rather anthropocentric view of things: it is we who decide the value of the environmental damage caused by the pollution, not the environment itself. It is therefore quite conceivable that the economically optimum level of pollution will lead to irreversible environmental damage. We would regard the extinction of the Californian condor as a disaster; the extinction of the smallpox virus, by contrast, is hailed as a great achievement. In fact, even if we so wish, we are not actually able to put a precise figure on the value of environmental damage. What is a species worth and more to the point, to whom? And who makes the judgment: you, me or the species itself?

The problem therefore remains: how do we decide what is a proper level of environmental damage? It is a recurring problem. The fixing of emission or environmental quality standards for certain pollutants, defining the conditions to be attached to an operating permit for an industrial installation, authorizing large-scale public works projects, protecting the habitat of endangered species: all involve a judgment as to what society regards as an acceptable, or unacceptable, level of environmental damage.

Many issues could have been chosen to illustrate this problem. But one has become a classic of its kind. In fact it might be thought of as a spicy issue, involving as it does the longest river in Western Europe and the need to dispose of 7 million tonnes of salt a year.

THE RIVER

The longest river in Western Europe is the Rhine. From its source in the Swiss Alps to its estuary on the Dutch North Sea coast, it is 1320 km long. The catchment area of 185,000 km^2 extends into no less than eight countries and one principality: Italy, Austria, Liechtenstein, Switzerland, France, the Federal Republic of Germany, Luxembourg, Belgium and the Netherlands.

At its source in Switzerland the Rhine is a glacial river. Its tributaries, however, such as the Neckar, the Main and the Moselle, are fed by rain. The combined effect is to give the Rhine the characteristics for much of the year of a rain-fed river, but in the drier months of summer and autumn those of a glacial river. This hydrological regime, resulting in a peak flow rate in early summer and low water levels in late autumn and winter, has been a major factor in the development of the river valley. In particular, the periodic floodings contributed to the evolution of various unique biotopes along the river, but have also, over the past 200 years, encouraged the construction of a large number of river management schemes in the interests of flood control, navigation and the generation of hydro-electicity.

The economic importance of the river is well known. Rotterdam, the world's largest port, is built on its estuary; one of Europe's biggest inland harbours is located at Duisburg; and a large number of major industrial centres such as Basel and the Ruhr conglomeration are scattered along its banks and tributaries. The river provides a means of transport – about 300 million tonnes of cargo are shipped along its

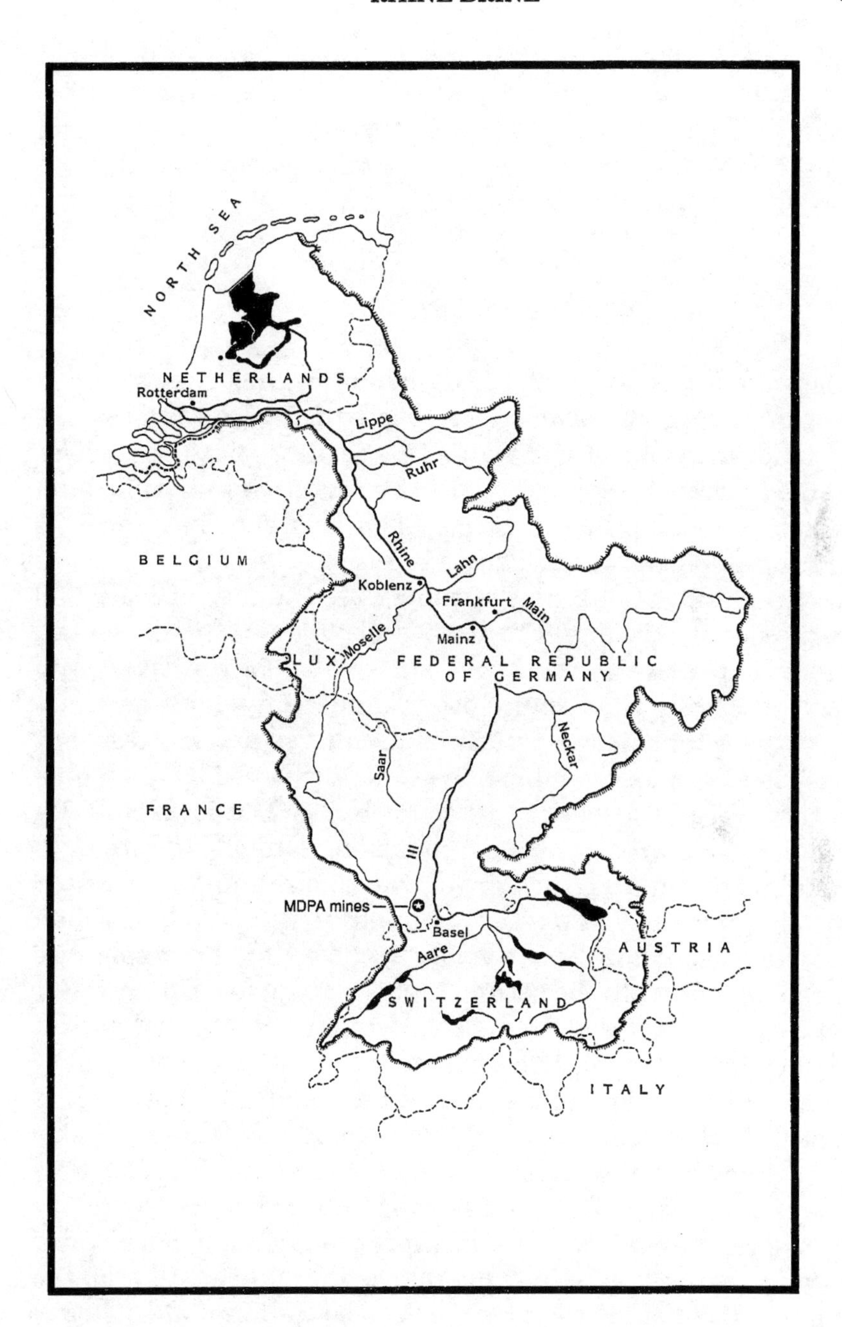

The Rhine catchment area

lower reaches each year. It also provides a plentiful supply of water for a wide variety of purposes: process water for manufacturing purposes, cooling water for industry and power stations, drinking water for 20 million people and irrigation water for agriculture.

THE SALT

The popular conception of the Rhine is that it is an open sewer. It might, however, be argued that this impression is inaccurate on two counts. In the first place, no sewer has ever had to accommodate the quantities of pollutants which the Rhine has to carry. Each year the river receives around 3000 tonnes of zinc, 1100 tonnes of oil, 700 tonnes of chromium, 500 tonnes of copper, 400 tonnes of lead, 350 tonnes of nickel and 150 tonnes of arsenic. In the second place, and more importantly, the quality of the river is in fact far better than many other industrial waterways; not that the quantities of polluting discharges are any less, but because the sheer volume of water transported by the Rhine – the average discharge into the North Sea is 2200 m^3 a second – ensures that any pollution is highly diluted. To take two standard indicators of water pollution, water tested for the biochemical oxygen demand at the Dutch–German border, just before the river branches into three separate streams, falls into the second highest of the five Dutch water quality classes, and the concentrations of the eight most important heavy metals are all below the strict basic quality standards laid down for Dutch surface waters. This is not to say, however, that there are no serious pollution problems associated with the Rhine. The river, particularly in its lower reaches, is such a dominant factor in the hydrology of the region that a general improvement in surface water quality is only possible by improving the quality of the Rhine. The bed of the river is also contaminated by heavy metals, again, especially in the lower reaches where the flow is more sluggish and suspended particulates tend to settle out.

Curiously, one of the most serious pollutants is sodium chloride – common salt. Natural processes ensure that salt is present in all fresh water in relatively low concentrations; in the case of the Rhine, the natural level of salt in the lower reaches of the river is about 20 mg/l. Over the past hundred years, however, salt has been discharged into the river in increasing quantities, so that the average concentration at the Dutch–German border is now nearly fifteen times greater than the natural level at about 290 mg/l. This represents a load of 620 kg of salt a second, 53 000 tonnes a day, 19 million tonnes a year.[1]

As might be expected, increasing the salinity of the Rhine to many times its natural level has a number of undesirable effects. When used for irrigation purposes in agriculture, the water causes lower yields and damages the quality of the agricultural products; industries which abstract water from the river for use in their manufacturing processes have had to upgrade their purification equipment; and drinking water supply companies which are dependent on the Rhine for their supplies of fresh water have had to invest in additional treatment plant in order to meet drinking water quality standards and to limit the corrosion caused by the salt to steel and cast-iron water mains. The problem is exacerbated by the large seasonal fluctuations in the salinity of the Rhine. During periods of low flow, such as late autumn, the average chloride concentration of 175 mg/l may well increase by 75 per cent, so that users of the water have to be prepared to cope with periodic peaks in the salinity. In 1985, for example, when a drought led to an exceptionally low flow, a chloride concentration at the Dutch–German border of 415 mg/l was measured.

---

[1]The salinity of water is usually expressed by reference to its chloride concentration. A concentration of 290 mg/l of salt is equivalent to a chloride concentration of 176 mg/l. This convention will be followed in the rest of the chapter.

The discharge of some 17 million tonnes of salt into the Rhine each year attests to the scale of the industrial activity along the river. Major chemical companies such as Hoechst, BASF, Ciba Geigy and Sandoz, along with many other industries, are all sited on the banks of the Rhine or its tributaries and contribute to the problem by discharging chlorides and sulphates. The coal mines in the Ruhr also discharge large quantities of chlorides. But by far the largest single source of salt in the Rhine – accounting for about 40 per cent of the total load – is the state-owned mining company *Mines de Potasse d'Alsace SA* (MDPA). Located near Mulhouse in the Alsace region of eastern France, MDPA mines sylvite as a raw material for artificial fertilizer. Sylvite is a mineral made up of about one-third potassium chloride and two-thirds sodium chloride. The potassium chloride is separated from the sodium chloride and sold as fertilizer. The sodium chloride – about 7 million tonnes per year – remains as waste.

PERMISSIVE PERMITS

MDPA started operating in 1913, since when the production capacity has steadily increased to the point where about 50 000 tonnes of sylvite is extracted each day. Until 1931 the waste salt was simply dumped in huge mounds near the mines. However, no precautionary measures were taken to prevent the infiltration of the salt into the soil, with the result that the groundwater in the region became seriously contaminated. An obvious alternative for disposing of the waste salt was provided by the Rhine, which conveniently passed about 15 km from the mines. The salt could simply be dissolved in water and discharged into the river through open channels. Thus it was that, on 24 April 1931, the prefect of the *département* Haut-Rhin granted the company a permit to dispose of the salt in the Rhine. From that day on the issue became an international *cause célèbre*.

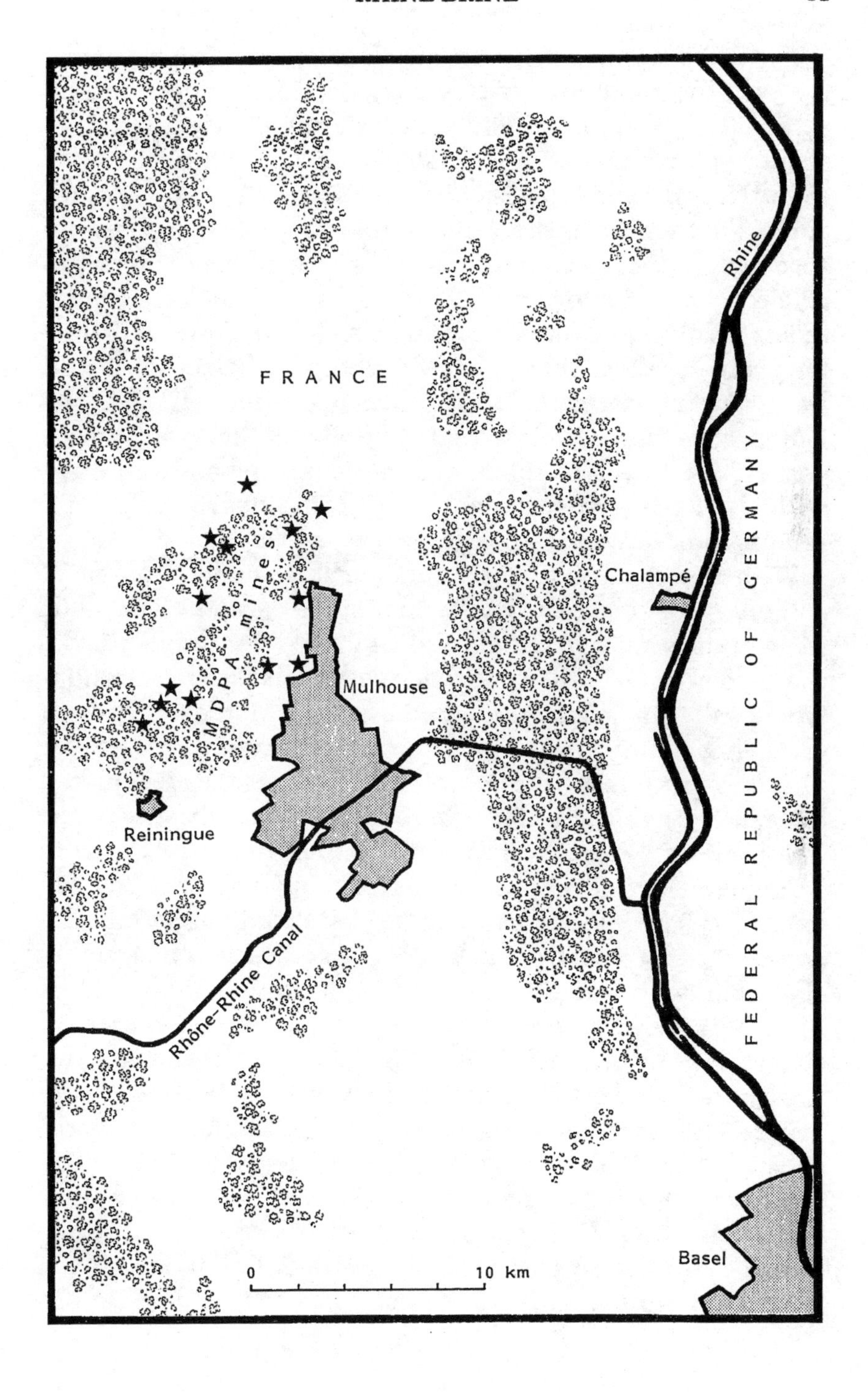
FRANCE
Rhine
FEDERAL REPUBLIC OF GERMANY
MDPA mines
Chalampé
Mulhouse
Reiningue
Rhône-Rhine Canal
Basel
0
10 km

The most crucial condition attached to the discharge permit was the requirement that the concentration of chloride in the Rhine immediately downstream of the discharge should not exceed 200 mg/l. This was about 125 mg/l higher than the average chloride concentration in the river at that point. Further down the watercourse at the Dutch–German border, the discharge had the effect of increasing the chloride concentration by over 50 mg/l. It was immediately clear to the drinking water companies which relied on the Rhine for their supplies that the new method of disposal would have serious consequences for their operations. Indeed, the Dutch government felt so strongly about the issue that a year later, in April 1932, it entered a diplomatic protest with the French government – but to no avail: the permit remained in force and the discharges continued.

With the exception of a sharp decline in the amount of salt discharged into the Rhine during the Second World War, little in the situation changed until the 1950s. Then, in 1955, the prefect in Haut-Rhin authorized an increase in the quantity of salt which MDPA could discharge. Moreover, instead of setting a limit on the chloride concentration in the river just downstream of the mines, the new permit designated Rees, near the Dutch–German border, as the reference point. The limit set – a maximum contribution from the mines of 88 mg/l in the total chloride concentration when the flow fell below 1700 m^3/s – represented a substantial increase in the amount of salt which could be discharged. This relaxation in the permit conditions was vital to the continued operations of MDPA. In fact production at the mines had increased so much over the previous years that, even allowing for the more generous permit, MDPA was obliged to build several reservoirs on an island in the Rhine just north of the mines. These provided the company with a buffer capacity sufficient to hold back part of the discharge until the weekend or during periods when the flow of the Rhine was too low to assimilate the salt. Regrettably the risk of saline infiltration was overlooked, with the result that the

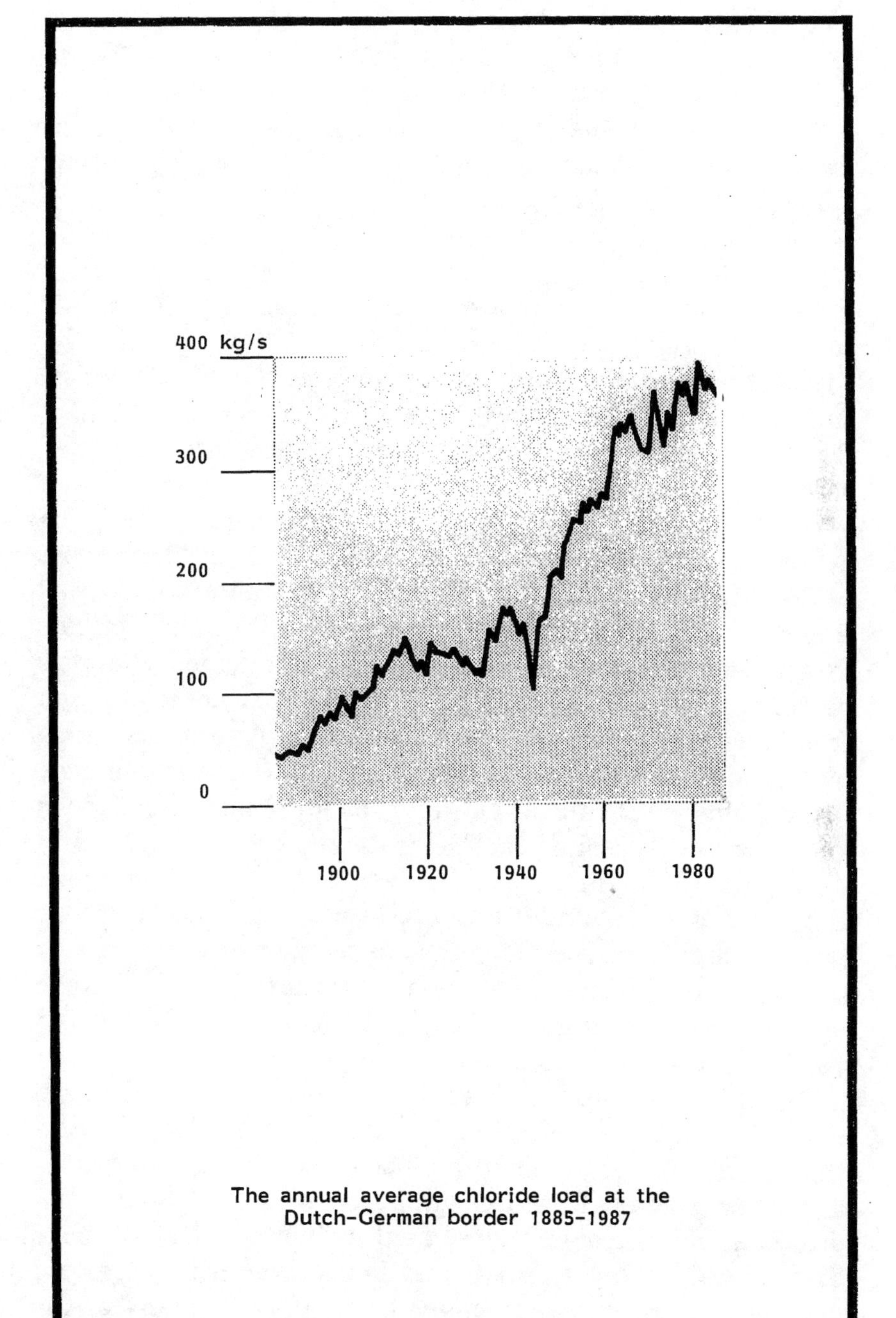

The annual average chloride load at the Dutch-German border 1885-1987

groundwater in the Alsace and Baden-Württemburg became contaminated. Under pressure from the government in the German state, the use of the reservoirs was ended in 1976, even though this inevitably caused greater fluctuations in the chloride concentration of the river. The island is now home to a nuclear power station.

These developments had been followed with increasing concern by those with an interest in a clean Rhine. Protests from other Rhine states and from the drinking water companies along the river had achieved little, so that by the early 1970s the time had come for a change of tactics. Almost simultaneously two new offensives were launched. Ironically, the first involved international litigation, the second international cooperation.

On 4 October 1974, three Dutch market gardeners who drew their irrigation water from the Rhine, with the support of the environmental group Reinwater Foundation, entered a civil action simultaneously in the district courts of The Hague and Rotterdam. The plaintiffs demanded that the discharges be declared illegal and claimed compensation from MDPA for the damage caused to their nursery crops by the high salt content of the irrigation water which they drew from the Rhine. If the market gardeners expected a speedy judicial resolution of the issue they were to be greatly disappointed. In the first place the court in The Hague took three years to conclude that it was not competent to judge the case and referred the claim to the Rotterdam court. It took another five months for the Rotterdam district court to agree to hear the claims jointly, so it was not until November 1977 that real progress could be made.

At about the same time as the market gardeners started their action, the International Commission for the Protection of the Rhine against Pollution, commonly known as the International Rhine Commission or IRC, began preparations for a formal agreement to control the chloride discharges. In fact the IRC had been established nearly a quarter of a century previously, in 1950, as a forum for the five riparian states – Switzerland, France, Luxembourg, the Federal Republic of Germany and the Netherlands (joined in 1976 by the European Community) – to discuss common problems associated with the Rhine. The body was given formal status in 1963 with the signing of the Berne Convention and charged with three tasks:

- the coordination of research into the nature, extent and origin of pollution and the interpretation of the results;
- the proposal to the signatory states of appropriate protection measures;
- the preparation of international treaties for the protection of the Rhine.

In order to facilitate the execution of these tasks, the IRC maintains a permanent scientific and technical secretariat at its headquarters in Koblenz.

In the 1960s the IRC had carried out several preliminary studies to examine various ways of reducing the chloride load of the Rhine. On the basis of these studies, agreement was reached between the Rhine states at the first Ministers Conference on the Pollution of the Rhine, held in The Hague in October 1972, that by 1 January 1975 and for a period of 10 years MDPA would hold back 260 kg/s of its chloride discharge by tipping it on nearby land. The costs of doing this would be met jointly by the Rhine states. This commitment, however, was not met. Instead, at the third Ministers' Conference in April 1976, France proposed the more permanent solution of injecting the chloride into a porous limestone formation 1500–2000 m below ground in the Alsace area. This alternative was accepted by the other countries and formed the basis for the first formal agreement between the Rhine states on the chloride problem: the Convention on the Protection of the Rhine against Pollution by Chlorides, signed in Bonn on 3 December 1976 by the governments of the five riparian states.

It is worth pausing at this point to note certain features of the agreement. In the preliminaries to the convention, the desire to limit the chloride concentration of the Rhine at the Dutch–German border to 200 mg/l was noted.[2] A series of measures were then laid down which were to contribute to achieving this goal. First, and most important, France agreed to reduce discharges of chlorides to the Rhine on its territory by 60 kg/s, measured as an annual average. This was to be achieved in phases. In the first phase the French government was to establish, as soon as possible after ratification of the convention, a facility to inject waste salt from the mines deep below ground for a period of 10 years at an initial rate of 20 kg/s. In a second phase and on the basis of the results of this operation, the French government was to take further measures in order to secure the total reduction of 60 kg/s by 1 January 1980, conditional on agreement on the technical aspects and the financing of the measures. In addition to these reductions, the IRC was also required, within four years of the convention coming into effect, to draw up proposals for a third phase of reductions aimed at further limiting the chloride load along the entire length of the Rhine.

Second, the five signatory states committed themselves to the "standstill principle" by taking all necessary measures to prevent any increase in the chloride load of the Rhine on their territories. A new chloride discharge or an increase in an existing discharge was therefore to be compensated by equivalent reductions elsewhere in the country.

Third, although financial responsibility for the deep-well injection was primarily to be borne by France, the other four countries agreed to make a fixed contribution to the

---

[2]To put this standard into perspective, the 1975 EC directive which fixed standards for the quality of surface water that is abstracted for drinking water supplies requires the member states of the EC to "endeavour to respect" a maximum chloride concentration of 200mg/l. The 1980 EC directive on the quality of drinking water defined a chloride concentration of 200mg/l as the level above which health effects "might occur" and laid down a guide level of 25mg/l. Dutch legislation itself lays down a maximum annual average limit value for chloride in drinking water of 150 mg/l.

estimated costs of FF 132 million ($30 million). The share of this contribution was fixed by the convention: Switzerland 6 per cent, Germany 30 per cent and the Netherlands 34 per cent, leaving 30 per cent to be paid for by France. This was the most remarkable feature of the agreement, and an awkward precedent which the signatory states did not care to be reminded of when other transboundary pollution issues arose in later years. The decision by the other countries to share the costs of deep-well injection in France was clearly determined by considerations of political expediency rather than the prescriptions of international law. They knew that they had a good case against France if the question was to be brought before a court of law, but they were also highly conscious of the fact that it would be many years before a final judgment would be made – and that no hard sanctions could be applied to enforce a positive verdict. If a blind eye were to be turned to the awkward breach of the "polluter pays principle", the acceptance of a share of the FF 132 million to ensure speedy action and to avoid a lengthy and politically acrimonious judicial battle could be regarded as a positive bargain.

SECOND THOUGHTS

Signing a convention is one thing, ratifying it is another. Because of the complexities of conducting negotiations between several countries, it is accepted practice for the respective foreign ministers to agree on the text of a convention and for the governments to sign the document before formal approval is sought from the national parliaments. This is not to say that consultations with parliamentary representatives do not take place – most governments will want to have an indication that support for a particular standpoint is likely to be forthcoming – but involving the elected assemblies during the detailed negotiations would inevitably cause excessive delay as all kinds of detailed considerations would be raised. Instead, the key role of parliaments in this process is to agree to ratify the convention, as it is only when an agreed number of the signatory states have deposited instruments of ratification that a convention will enter into force.

In the case of the Rhine Chlorides Convention, Article 14 provided that the agreement would enter into force on the first day of the second month following the deposition of the final instrument of ratification. The parliaments of Switzerland, Germany, Luxembourg and the Netherlands duly approved ratification, leaving the final word with France. Despite some vociferous resistance in the Alsace region to the idea of deep-well injection – due to the costs and the fear that groundwater in the locality would be contaminated – approval was expected without serious problems, particularly as President Giscard d'Estaing had publicly declared that the convention should be ratified. However, on 3 December 1979, when the National Assembly was due to decide the date on which it would debate the proposal, the government unexpectedly announced that it was withdrawing the bill on the grounds that there was unlikely to be a majority in support of ratification. Intensive lobbying by a number of influential Gaullist deputies, who had recently flexed their political muscles by securing a defeat of the 1980 budget, seemed to have persuaded the government that it might suffer further damage if the issue was debated in parliament.

To the other Rhine states the decision was clearly a politically expedient response to a highly contentious issue. All four condemned the French move, the Dutch government even going so far as to recall its ambassador in Paris to The Hague "for consultations". The development also raised the awkward question of what to do about the FF 45 million ($10 million) already paid to the French government under the convention as an advance to help cover the costs of deep-well injection. The French government suggested that it reimburse the payments, but the other countries refused the offer, preferring to leave the money where it was as a cogent reminder of their commitment to the convention.

A new round of talks between the five countries was hastily arranged, but little progress was made. Quite the opposite, for the possibility of a speedy resolution to the issue became even more remote when the French delegation argued that deep-well injection was not, after all, the most suitable way of disposing of the waste salt and that other options should be evaluated, thereby opening up the possibility of deferring action for a lengthy period. Chief amongst these options was the construction of a salt works, which would turn the waste into a saleable product. To the other delegations the salt works was a non-starter. The idea, which was by no means new, would take many years to be realized and was far more expensive than deep-well injection, costing about FF 320 million ($80 million) so that France requested proportionately greater financial contributions from the other Rhine states plus a contribution towards the operating costs. It further suffered from the overwhelming political objection that the European market for salt was not only saturated but dominated by a small number of companies, including BASF in Germany and Akzo in the Netherlands. The German and Dutch governments were not exactly keen to help finance the establishment of a new competitor, and became even less so when the French suggested that the other Rhine states should agree to take a fixed proportion of the output produced by the works.

Complicating the issue even further, the French delegation tossed two further ideas onto the negotiating table. It was suggested that Germany, the second biggest discharger of chlorides into the Rhine, reduce its own discharges, even though it was under no obligation to do so under the 1976 convention, and further that the discharges from the many smaller industries along the Rhine should also be reduced. It came as no surprise to anyone when the other delegations reacted coolly to these various proposals. The communiqué issued at the end of the meeting recorded soberly that agreement was reached for the IRC to meet again the following month when France would present more detailed proposals for discussion. No mention was made of the fact that by then 1 January 1980 would have passed – the date when, according to Article 2(3) of the Rhine Chlorides Convention, MDPA was to have reduced its discharges to the river by 60 kg/s.

The meeting, held in Brussels on 28 January 1980, proved fruitless. Instead of presenting concrete proposals for reducing the discharges, the French delegation reiterated its view that the construction of a salt works was the most appropriate option. This would need to produce a million tonnes of salt a year in order to secure the agreed first-phase reduction of 20 kg/s. Further, not only should Switzerland, Germany and the Netherlands contribute to the costs of the salt works on the same basis as was agreed for the deep-well injection (6 per cent, 30 per cent and 34 per cent respectively), but they were also expected to take a fixed proportion of the output. Objecting that the market could not absorb such large quantities of salt, the other delegations suggested other, and to them more practical, disposal options such as further dumping on land close to the mines, returning the salt to worked-out parts of the mines, or cleaning the salt and shipping it to the soda works in the nearby Lorraine region. Once more, however, the only agreement which could be reached was to study the various options further and to meet again, this time in May.

By now the writing was clearly on the wall, despite direct talks in March between the French and the Dutch Prime Ministers, Raymond Barre and Dries van Agt. The problems were highlighted at the May meeting. France again insisted that a salt works was the only feasible way of dealing with the problem and the other delegations duly reiterated their objections to the option. All that could be agreed was that the Chloride Coordination Committee, a working group comprised of experts from the Rhine states and under

Swiss chairmanship, should draw up an inventory of the various options for discussion in a plenary meeting of the IRC in June. Given the entrenched positions of the delegations, it was not expected that this approach would lead to a resolution of the differences. The expectations proved justified, and at the plenary meeting on 9–10 June the IRC could only agree to refer the matter back to the delegations for further study.

THIRD THOUGHTS

By this time – nearly six years after the action had been brought – some progress had been made by the Rotterdam District Court in the action brought by the market gardeners against MDPA. On 28 April 1980 the court ordered that an international commission of experts be appointed to report on the extent to which the discharges by MDPA contributed to the salinity of the irrigation water used by the market gardeners. This report would then form the basis for a judgment by the court. Three experts, one each from France, Belgium and the Netherlands, were appointed in November 1980 and set to work on a meticulous study of the complex material. Clearly, the judicial option was not going to provide the parties with a speedy resolution of the issue.

There seemed to be little cause for hope. But, at the next meeting of the national delegations, on 9 December, a certain degree of flexibility on the part of the French delegation opened up the prospect of an agreement. It was agreed that a working group should study the feasibility of shipping a million tonnes of waste salt each year down the Rhine for dumping in the North Sea and to report on the practical aspects of the proposal to the next Ministers' Conference, due to be held on 26 January 1981. It was certainly not the cheapest option on offer – the plan would require the permanent services of 20 barges – but at least it offered the prospect of an acceptable compromise.

In fact the fifth Ministers' Conference made progress in several respects. In addition to retaining deep-well injection as an option, it was also decided to carry out a detailed technical and economic analysis of the proposal for shipping the salt down-river to the mouth of the Rhine and then discharging it into the North Sea through a pipeline. The barge operators were understandably enthusiastic about the idea, but two serious problems were pointed out. In the first place the operation was expensive: at a cost of FF 120 million ($27 million) a year it was five times more expensive than deep-well injection. In fact the costs would be so high that it might prove cheaper for the Dutch government to compensate its market gardeners for the damage caused by the salt than to pay its share of the shipping costs. Secondly, the salt was contaminated with potassium ferrocyanide, a substance added to the waste salt in order to reduce its viscosity and thereby improve its flow qualities. Both Germany and the Netherlands made it immediately clear that it was unacceptable to dump the waste salt in the North Sea unless the potassium ferrocyanide was first removed. The delegates also agreed that the French government would study the feasibility of constructing a pipeline to transport the waste salt more than 100 km to the soda industry in the Lorraine where it could be used as raw material. Little was expected of this option, however, given the high costs – FF 60 million ($14 million) a year – and the intrusion which such a pipeline would cause.

Perhaps as important as the decision to assess further the three options was the fact that the possibility of depositing the waste salt in the exhausted mines was finally rejected. (Rumour had it that France had turned its eye to the mines as a suitable site for the storage of radioactive wastes from its nuclear power stations.) The French government also accepted the objections of the other states to the construction of a salt works – against the wishes of the 6000 employees of MDPA who promptly went on strike and persuaded the Minister of Environment, Michel d'Ornano, to promise a study into the economic feasibility of a French-financed salt works.

In the meantime the lawyers had once again been busy. The reason for their activity was the decision in December 1980 of the prefect of the *département* Haut-Rhin to extend MDPA's discharge permit, which was about to expire, for 12 months. Ten Dutch local authorities, water authorities and drinking water companies appealed against this decision to the Administrative Court in Strasbourg on the grounds that the prefect had failed to comply with the legally prescribed procedures and that he had acted in contravention of international law by authorizing an activity which caused damage in another country. The reaction of the prefect to this appeal was interesting to say the least. A month later he duly repealed the new permit, but immediately granted a new authorization which allowed MDPA to continue discharging chlorides at an average rate of 130 kg/s until 31 December 1990. The effect of this move was to nullify the administrative action against the old permit, obliging the Dutch objectors to initiate a new appeal.

But it was not only at the local level that procedural delays arose. In June 1981 the French government requested that the next ministers' conference, planned for the next month, be deferred until after the elections to the National Assembly in September. Accepting the political reality that the French parliament would be unwilling to approve a contentious proposal before the elections, the other Rhine states reluctantly agreed to a new date in November. This delay would at least have the merit of allowing the ministers sufficient time to assess the results of the studies into alternative methods of disposing of the waste salt. In the event these studies concluded that all three proposals – deep-well injection, shipment down the Rhine and transport by pipeline to the Lorraine – were technically feasible. The most important differences concerned the costs likely to be incurred. Deep-well injection at a rate of 20 kg/s was estimated to cost a total of FF 210 million ($40 million) over

10 years, transport by pipeline at the same rate about FF 315 million ($60 million) and shipment down-river to the North Sea at least FF 1 billion ($ 190 million).

Clearly, the ministers had something to get their teeth into in November. The new French Minister of Environment, Michel Crépeau, nevertheless decided that a rather different menu should be served up at the meeting. Following a visit to the mines in October, he declared that the issue should be tackled in a different way, even if that meant renegotiating the Rhine Chlorides Convention. This was another way of saying that France still thought a salt works was a good idea after all. The French government knew that the other Rhine states would insist that a salt works was still as bad an idea as it always had been – which they duly did – but as a negotiating ploy the approach had a measure of success. For although the conference concluded that deep-well injection was the most practical option for disposing of the waste salt, it was also agreed that about 30 per cent of the agreed initial reduction in the chloride discharge could be secured through the construction of a small salt works, provided that this was financed by France. Moreover, deep-well injection would only commence once a commission of experts, drawn from countries not party to the convention, had carried out a study into the environmental impact of deep-well injection in the Alsace and then convinced the local population that the technique did not pose a serious risk of contaminating groundwater reserves. No agreement was reached on exactly how long this process would last, although Michel Crépeau did his best to ease the worried expressions on the faces of the other ministers by suggesting that it might take only six months.

To no-one's surprise it turned out to be a year before the commission's findings were released. In November 1982 the experts reported that deep-well injection posed no serious threat to local groundwater quality. There was, however, a problem. The French government now suggested that the salt might be injected near Chalampé on the German border rather than in the vicinity of Reiningue, to the west of Mulhouse, as specified in the annex to the Rhine Chlorides Convention. What the German government thought about that remained to be seen. Nevertheless, the commission's positive report encouraged the new French Prime Minister, Pierre Mauroy, to announce that his government would, at long last, submit the Rhine Chlorides Convention to the National Assembly for formal approval.

Once again, however, things were not quite as simple as they seemed. In the first place, although a positive decision by the French parliament would mean that all five Rhine states would have ratified the convention, it had now become necessary to make a number of amendments to the text. The date by which the discharges were to be reduced by 60 kg/s had, for example, been fixed in the original convention at 1 January 1980. The location where the waste salt was to be injected into the ground – to the southwest of Mulhouse – had also been specified in the annex to the convention. French ratification was therefore to be conditional on the IRC reaching agreement on these "details". The amendments could then be formalized through an exchange of letters between France and the other four Rhine states (although these letters would again need to be ratified by the respective parliaments). The preparations for ratification might, therefore, take a little time.

In fact they took a year, the exchange of letters being completed in May 1983. Then, on 7 October 1983 and with the explicit support of the new president, François Mitterrand, a bill to approve formal ratification of the convention was submitted to the National Assembly. Although less than 30 of the 500 members were present in the chamber, the fierce opposition from the Alsace representatives guaranteed a colourful debate that lasted for more than five hours. Why was no action being taken against dischargers of chlorides in other countries? What of the fact that Germany and the Netherlands discharged far greater quantities of mercury, cadmium, arsenic and lead into the Rhine than France? And why should the other European producers of salt be permitted to obstruct the construction of a new salt works in the Alsace? In spite of these arguments, the final vote, in favour of ratification, was decisive: 275 votes for, 152 against. A month later the Senate followed the National Assembly's example and also approved the bill. After seven years of resistance, the French had finally ratified the convention.

BACK IN COURT

Progress could also be reported in one of the two legal actions against the discharges. After considering the matter for two years, the Administrative Court in Strasbourg finally came to the conclusion, in July 1983, that the permit granted to MDPA in 1981 which authorized a chloride discharge of 130 kg/s until 1990 should be annulled on the grounds that the discharge demonstrably caused damage to downstream users of the Rhine. The clear implication of the judgment was that any new permit granted by the prefect would have to take the interests of other riparian water users into account in the limits to be imposed on discharges of chlorides. The response of MDPA was immediate: the company announced that it would appeal against the ruling to the Council of State, and with good reason, for, judging by previous experience, it would be several more years before a final judgment was made. The response of the local prefect was equally resolute: a week later he granted yet another permit which was virtually identical to the permit which had been annulled by the court. The reaction of the Dutch public authorities was predictable: within a week a further appeal against the new permit was lodged with the Administrative Court in Strasbourg.

The activity in the original court action initiated by the market gardeners in 1974 had been somewhat less eventful, although progress was nevertheless discernible. The three experts appointed in 1980 by the Rotterdam District Court finally reported in late 1983 on the extent to which the discharges of chlorides from the mines contributed to the damage suffered by the nurseries. Their main conclusion was that the discharges contributed to between eight and 17 per cent of the salinity of the water used by the market gardeners for irrigating their crops. Both parties, however, saw this finding as vindicating their respective positions: to MDPA it demonstrated that other sources of salt in the Rhine greatly exceeded the contribution from the mines and that it would be unfair if their company alone should be penalized for the damage; to the market gardeners the finding was hard evidence that the discharges from the mines were responsible for enough additional contamination to require substantial investment in desalination equipment. On 16 December 1983 the court displayed its wisdom by finding that the market gardeners did indeed suffer unreasonable damage as result of the discharges from the mines and ordered MDPA to pay damages for the loss suffered since 4 October 1974, but only equivalent to the proportion of the saline content of the irrigation water for which the mines were responsible. It seemed a logical verdict, but MDPA, fully aware of the far-reaching implications of the judgment, drew its own logical conclusion and duly appealed to the Court of Appeal in The Hague. The lawyers were going to be in business for a few more years.

In fact they would be working overtime, for on 7 October 1983, the ten Dutch public authorities which were already involved in two court actions against the discharges decided to make a criminal complaint against MDPA to the District Court in Mulhouse. The accusation was that, by continuing to discharge chlorides at the rate authorized under the permit which had been annulled by the Administrative Court, the company was in criminal breach of the law. The first task of the court, however, was to decide whether it was competent to hear the case, which it duly confirmed 16 months later on 16 February 1985. However, an appeal against this decision by the public prosecuter was upheld by the Court of Appeal in Colmar four months later, although the Dutch public authorities then appealed in their turn to the High Court. It would be another two years before that judgment was made.

If the lawyers were enjoying themselves puzzling over the finer points of French and international law, at least the ratification of the Rhine Chlorides Convention by the French government now held the promise that some reduction in the discharges would shortly be secured. Even so, it was not until 5 July 1985 that the last of the Rhine states – ironically the Netherlands – completed the ratification procedure for the amendments to the convention. Under the amended text, France was allowed 18 months from the date on which the convention came into effect (until 5 January 1987) to secure the first-phase reduction of 20 kg/s of chlorides, and a further two years (until 5 January 1989) to reduce the discharges by the total amount of 60 kg/s.

Complications persisted, however, with the issue of deep-well injection. This method of disposing of the waste salt was explicitly provided for in the convention, which France had now ratified. But the convention also included an escape clause: the French government was authorized to suspend deep-well injection if the process posed a serious hazard to the environment, in particular to groundwater quality. As a final decision had yet to be taken on where the deep-well injection would take place, it was by no means certain that this method of disposal would indeed prove to be acceptable; it had not been possible to start drilling a trial borehole to test whether the geology of the area near Chalampé was suitable for deep-well injection until May 1985, as the site had been occupied for a year by local protesters. Preliminary results were expected in early 1986. The French Foreign Minister, Roland Dumas, had nevertheless given an assurance in July 1985 that France would take the measures necessary to implement the convention by 1 January 1987. At the same time it was clear to experts from the other Rhine states that, should Chalampé be chosen as the most suitable location, it would be impossible to meet the deadline of 1 January 1987 as Germany had already warned that objections would be made because of the risk that the groundwater in Baden-Württemberg, just a few hundred metres away on the other side of the Rhine, would be contaminated.

All that could be done was to wait for the results of the trial injections. Again, however, the prefect of Haut-Rhin succeeded in breaking the tension by unexpectedly issuing, on 5 September 1985, yet another new permit to MDPA. The new permit once again authorized discharges at the existing levels and the Dutch public authorities once again served notice of appeal at the administrative court in Strasbourg, making it the fifth concurrent court action on the legality of the discharges.

Meanwhile a verdict was being prepared in the second of the court cases, that involving the appeal by MDPA against the judgment of the Administrative Court in Strasbourg annulling the permits granted in December 1980 and March 1981. It was a notable judgment for, on 18 April 1986, the Council of State confirmed the annullment of the permit granted in March 1981, but found that the previous permit, granted by the prefect of Haut-Rhin in December 1980 and withdrawn by him in March 1981, had in fact been valid. The ruling was of little more than academic interest, however, given that the prefect had subsequently granted two more discharge permits to MDPA, both of which were the subject of appeals by the Dutch authorities.

The Dutch market gardeners were making rather better progress with their own action against the discharges. To be sure, they had been pursuing the matter since 1974 but they were ever optimistic of success. Further encouragement came in September 1986 when the Court of Appeal in The Hague, in its ruling on the appeal by MDPA against the 1983 judgment, confirmed that the discharges were unlawful. However, the court based this ruling on Dutch rather than international law, although as the Member States of the European Community mutually recognize judgments made in the other countries the practical effect was the same – or would have been the same, except that MDPA moved an action in the High Court to dismiss the verdict. The market gardeners responded by moving that the High Court dismiss the appeal by MDPA, with the result that everyone had to wait another two years for the final judgment.

BACK TO THE DRAWING BOARD

By now, developments elsewhere had diverted attention away from the courtrooms. In February had come the first news that the trial injections of waste salt near Chalampé were indicating that the geology of the area was unsuitable for deep-well injection. In particular, the porosity of the strata at a depth of 1600 m seemed to be insufficient to absorb the waste salt at the necessary rate. A commission of experts was studying the results and would shortly submit its recommendations to the government, but with elections due on 16 March the more cynical observers were resigning themselves once more to further delay and placed their hope instead on the next meeting of the IRC on 2 June.

This proved to be an interesting event, not least because six days before the meeting the new French Prime Minister, Jacques Chirac, telephoned the Dutch Prime Minister, Ruud Lubbers, to inform him that France wished to abandon the option of disposing of the waste salt by deep-well injection: the geology of the area around Chalampé was not suitable and strong resistance to the idea amongst the population in the Alsace persisted. There was no question, however, of France reneging on its commitment to reduce the discharges by 20 kg/s before the deadline of 5 January 1987. Strictly speaking, since there were no technical objections to injecting the salt at the original location of Reiningue (which had been assessed favourably by the experts), the other Rhine states could have insisted that there was still no good reason not to proceed with deep-well injection. Indeed, this possibility was raised during the IRC meeting after the leader of the French delegation had read out a somewhat equivocal statement proposing that an alternative method for reducing the discharges be devised and that an expert committee should be appointed to report to the French government by 10 October on the feasibility of the alternative methods. He concluded by announcing that he had no mandate to discuss the matter further. Pressure from the other countries nevertheless led to the French representative securing authorization direct from Jacques Chirac to confirm that the agreed reduction would be secured by 5 January 1987, though the delegation was still under instructions not to participate in any discussion. The Dutch delegation suggested that an extraordinary meeting of the IRC be called at an appropriate time in order to discuss any new proposal by the French government for an alternative method, although it did not go unremarked by the other participants that all other conceivable alternatives had already been rejected.

It was not until 6 October, with the deadline just three months away, that the French Prime Minister, in a meeting with the Dutch Prime Minister and Minister for Foreign Affairs, set out the new proposal. The reduction of 20 kg/s would be achieved in two ways: 750 000 tonnes of salt a year, equivalent to a discharge of 15 kg/s, would be deposited on land in the vicinity of the mines for a period of ten years (the duration of the Rhine Chlorides Convention) and the remainder would be achieved through a reduction in the discharges of chlorides from the soda works in the Lorraine – not from the MDPA mines – and enforced through attaching a special condition to the discharge permit. Conveniently, the latter reduction had already been achieved as a result of a reduction in the output from the works.

It seemed a deceptively simple way of complying with the convention, given that the problem had seemed intractable for so long. In fact, to the more seasoned observers the proposal brought the entire affair full circle: it was a method which had first been suggested in 1972 but rejected due to objections by the local population. The proposal had been resuscitated on several occasions but had been opposed by the French government because of the reputedly high costs of preparing the sites. Curiously, as the proposal now stood these costs would involve little more than sealing the surface of the depots with a layer of asphalt in order to prevent the salt from percolating into the groundwater and then ensuring that the saline run-off from the site was collected. The practice was nevertheless decidedly more complicated than the theory, although for reasons other than the cost. In the first place, appropriate sites on which to deposit the salt had to be found. It was therefore proposed to expand an existing depot where the salt spread on the region's roads during the winter was stored; extending the area of this depot by 8 ha would accommodate a further 250 000 tonnes of salt each year. The remaining 500 000 tonnes per year would have to be deposited on a new site, though this would involve buying several holdings from local farmers which might only be possible through the use of compulsory purchase procedures. The surface of both sites would then have to be sealed with a layer of asphalt. It was also obvious that the proposal only offered a temporary way out of the problem: some longer-term destination for the 7.5 million tonnes of salt which would be held back over ten years would still have to be found.

The new proposal was discussed at an extraordinary meeting of the IRC in Colmar, near the mines, on 3 and 4 November. Questioned on the deadline of 5 January 1987, the French delegation assured the other countries that a start could be made by that date in storing the necessary quantities of waste salt at the existing depot. A number of possibilities were raised for the final disposal of the salt: using it on roads during the winter; as a raw material for industrial purposes; or it could be transported from the Alsace to Marseilles through a disused pipeline and dumped in the Mediterranean Sea. The Dutch and German delegations were quick to point out that France should not start entertaining any ideas of gradually dumping the retained salt into the Rhine at a later date.

A month later, on 11 December, the IRC met in Brussels to consider formally whether the proposed method met the requirements of the Rhine Chlorides Convention. The result was a declaration, signed by all five parties, which confirmed that the proposal was indeed permissible provided that: the discharge of chlorides into the Rhine from French territory was equal to that required by the convention; that no further financial contributions need be made by the other four states; and that the contributions already paid would only be used to finance the measures necessary to reduce the discharges to the river.

Exactly on time, on Monday 5 January 1987, MDPA started storing waste salt at the existing depot near the mines. The result was a reduction in the discharge of chlorides by an average of 15 kg/s. Together with the reduction at the soda works in the Lorraine, the measure succeeded in reducing the salinity of the Rhine by 6 per cent.

DÉJÀ VU

The question now arose of how the next step, of a further reduction of 40 kg/s was to be achieved by 5 January 1989. The first indications suggested that a combination of measures might be chosen, although it was by no means certain what these measures might be. Only in January 1988, when the French government circulated an expert report on the various alternatives, did the options become clear. The choice had been narrowed down to two alternatives: expanding the existing storage facilities to accommodate a further 2 million tonnes a year or transporting the waste salt to the North Sea for dumping. A final decision would be made in June after a more detailed assessment of the cost and technical feasibility. The choice, when it was announced in June, was hardly a surprise. It was proposed to continue the current practice of storing the waste salt in nearby depots and to expand the facilities to accommodate the substantially greater quantities of salt. All that was now necessary was to gain the official blessing of the other Rhine states at the ninth Ministers' Conference on 11 October. Again, given the history of the problem, observers could be forgiven for wondering why it had taken so long to resolve the issue.

Before then, however, interesting developments were reported in two of the court actions against the discharges. First, on 17 May, after a High Court judgment in June 1987, the Court of Appeal in Paris finally upheld the appeal by the ten Dutch public authorities against the ruling of the court in Colmar not to hear their complaint that the continuing discharges constituted a criminal offence. The court also found that the discharges were in breach of the law, thereby providing the Dutch plaintiffs with grounds for claiming damages. This they duly did, entering their claim at the District Court in Mulhouse. On 2 September the court recognized the claim and, pending the assessment of the damages suffered, ordered MDPA to pay two of the plaintiffs provisional damages of FF 2 million ($315 000). MDPA naturally appealed against this judgment, and with success, for on 7 November the Court of Appeal in Colmar set aside the verdict of the District Court. The equally understandable response of the Dutch authorities was to appeal against that decision to the Court of Appeal in Paris.

The second, and more important development, came in the 14-year-long action by the three Dutch market gardeners. On 24 September 1988 the High Court declared that MDPA should pay damages to the gardeners as compensation for the losses which they were suffering as a result of the discharges from the mines. On the same day the market gardeners and MDPA jointly announced that they had reached agreement on a financial settlement. MDPA would pay the ten cooperatives which represented 95 per cent of all the market gardeners in the Westland area a sum of NFl 3.75 million ($1.7 million). In return, the cooperatives agreed not to initiate any further court actions against MDPA on behalf of their other members. It was a crucial point for the company, for the judgment had established the principle that a polluter is responsible for the damage which is caused in another country, even where the actions comply with national law or are authorized by a permit. On this basis many of the other market gardeners in the Westland area could have started their own civil actions against MDPA. A large number of such actions would almost certainly have had serious financial consequences for the company and forced its closure. Instead the settlement represented a once-only payment that permitted the mines to continue discharging the waste salt at the existing rate.[3]

DUTCH COURAGE

Throughout the entire affair it had been the Dutch who had campaigned most actively, both in the IRC and the courts, for a reduction in the discharges of salt into the Rhine. That was to be expected. The Netherlands lay at the end of the river and the country was the unwilling recipient of all the pollution which the other Rhine states chose to discharge; it was therefore the greatest beneficiary of any action taken to clean up the Rhine. But the agreement formalized in the Rhine Chlorides Convention also meant that the Netherlands was the single largest contributor to the measures taken in France to reduce the discharges of

[3]It should nevertheless be noted that the judgment has far-reaching consequences for transboundary water pollution by clearly establishing the principle that polluters can be held liable for damage caused by their discharges. Specifically, downstream users "may in principle expect that the river should not be extremely polluted by extensive discharges".

waste salt, paying 34 per cent of the costs. For the first-phase measures, this contribution had been fixed at FF 45 million. No limit had been fixed on the costs of the second-phase reduction of 40kg/s, but the costs were to be distributed between the five Rhine states in the same ratio as the first-phase measures. In late 1988, therefore, it was still uncertain exactly how much each of the countries would have to pay to secure the further reduction.

This question had been exercising the minds at the Dutch Ministry of Public Works for some time. As the authority responsible for water management in the Netherlands, it was this ministry which would eventually have to meet 34 per cent of the bill for the further measures. It had noted that the costs of storing a further 40 kg/s of waste salt in special depots had been estimated by the French government at FF 816 300 000 ($140 million), inferring a Dutch contribution of FF 277 542 000 ($48 million). This financial burden clearly weighed heavily on the mind of the Minister for Public Works, Neelie Smit-Kroes. In particular, she had asked herself the politically loaded question of whether, in terms of pollution reduction, there might be a better way of spending the money.

She gave her answer to this question on 11 October 1988 at the ninth Rhine Ministers' Conference in Bonn. Unfortunately, not only had the other delegations not expected her to ask the question, they were certainly not expecting the answer that she gave. Her conclusion, simply stated, was that she was not prepared to spend nearly FF 300 million on measures which would only succeed in reducing the total chloride load of the Rhine in the Netherlands from 350 kg/s to 310 kg/s. But what, then, of the convention? "I'm not revoking the convention, I'm just not cooperating in its implementation", was her reply.

It was a remarkable development. But the reasoning of Smit-Kroes was largely a consequence of two important developments. In the first place, the Dutch government was becoming increasingly concerned about the impact of other forms of Rhine pollution, particularly toxic substances, heavy metals and phosphates; the damage caused by these pollutants was regarded as far more serious than that attributable to salt. Secondly, the French government was now formally proposing, as part of the package of measures to secure further reductions in the discharges, that the millions of tonnes of salt which were to be stored alongside the mines should be dumped into the river after the Rhine Chlorides Convention had expired in 1998. In fact the costs of dumping the salt in the Rhine would account for about 40 per cent of the total costs of the measures – which the other Rhine states would also be paying for. Fully aware of the sensitivity of this proposal, the French government had decided to sweeten the pill by suggesting (though not formally proposing) that production at the mines could be ended, or at least decreased, so that there need be no increase in the rate of discharge after 1998.

All that could be done after the meeting had settled down was to note that the French proposal did not enjoy the support of all parties and that, under these circumstances, it would not be possible to implement the second phase of the reductions by 5 January 1989 as provided for in the convention. It was agreed to confirm this delay through an exchange of letters and to ask the IRC to consider what steps should be taken in the light of the Dutch standpoint, deferring any decision until the next Ministers' Conference in 1989. The legal implications, for example, were far from certain. Was the French government within its rights to dump the waste salt into the Rhine after the convention had expired in 1998? Could the Netherlands effectively veto that part of the convention which required France to make a further reduction in the discharges by withholding its agreement on the method to be adopted? If one of the parties withdrew its commitment to the convention, would the obligation on France to continue with the first-phase reductions until 1998 still be valid? The political implications were equally uncertain. Would the other Rhine states, and particularly France, respond by refusing to cooperate in reducing the discharges of those substances about which Smit-Kroes was concerned?

In economic terms, Smit-Kroes's argument was a rational approach to the difficulties of the situation. Political rationality, however, operates under a different logic. The first to emphasize this was the German Minister of Environment, Klaus Töpfer. He made it clear that, as far as he was concerned, the Rhine issue could not be resolved by reducing it to a choice between different categories of pollution abatement, but only by ensuring that all forms of pollution were reduced to acceptable levels. The environmental groups were also less than happy. The Reinwater Foundation felt that, instead of giving up on the convention, it would have been far more credible, given the traditional Dutch standpoint on the issue, to press hard for an agreement whereby the salt which was stored was not dumped in the Rhine after 1998. Next in line was the Dutch Rhine Municipalities Steering Group, which came to the conclusion that tackling one form of pollution did not exclude action to reduce discharges of other pollutants. Then the lower chamber of the Dutch Parliament, in a debate with the minister, criticized both the tactic, fearing that it would rebound on the Netherlands in the IRC, and the fact that the initiative was taken at a ministers' conference and thereby ensured a high level of negative publicity. A motion was even adopted by a substantial majority urging the minister to press for the reductions prescribed by the convention. Finally, the Association of Drinking Water Companies was surprised to learn that the minister who was directly responsible for water quality had decided without consultation that salt was suddenly no longer a serious pollutant, even though the companies were faced with investments totalling around NFl 300 million ($140 million) in order to comply with the drinking water quality standards of 120 mg/l of sodium and 150 mg/l of chloride.

By the summer of 1989 it was becoming apparent that a

slightly less aggressive stance was being taken by Smit-Kroes, and in June the Dutch Minister of Environment, Ed Nijpels, revealed that a "good and sensible compromise" had been reached with the French government. Apparently the compromise was that the second-phase reduction need only be a further 20 kg/s instead of 40 kg/s, but that the rate of discharge would be adjusted to take into account the rate of flow of the Rhine: the greater the flow, the greater the permissible discharge. The broad effect would be to allow the mines to discharge larger quantities of salt in the spring when flow rates were highest and less in the autumn. Smit-Kroes, however, was less than pleased with her colleague's statements, for the very good reason that the German government was opposed to any renegotiation of the convention (fearing, perhaps, that the chloride discharges in Germany might be included in any new deal) and because discussions between the Dutch and the German governments on the proposal had yet to take place.

The prospects for progress became clear at the tenth Ministers' Conference, held in Brussels on 30 November 1989. The conference was remarkable in one respect at least: the discussions on further control measures focused on a joint proposal submitted by the French and Dutch governments – an unlikely coalition given the history of the issue. The proposal comprised three main elements. First, during periods when the chloride concentration of the Rhine at the Dutch–German border "substantially" exceeded 200 mg/l, MDPA would reduce its discharges by an appropriate amount (although due to the nine-day transit time between the Alsace and the Dutch border, it would be necessary to process information from several monitoring points along the river and use computer projections to estimate in advance when this limit was likely to be exceeded). Second, should the discharges be reduced as a result of a fall in the output of the mines, the salt stored in the depots could be dumped in the Rhine in an "ecologically responsible way". Finally, the Netherlands would take remedial measures to limit the effects of the discharges on its drinking water supplies. These measures would aim to reduce the input of saline water from reclaimed land which flowed into Lake IJssel – a large freshwater body supplied by the Rhine and an important source of drinking water in the Netherlands – by pumping the water into the North Sea. The total costs of the measures for the duration of the convention were estimated at FF 500 million ($80 million) and the share paid by each country would be as fixed by the convention: the Netherlands 34 per cent, France and Germany 30 per cent each and Switzerland 6 per cent.

Although the German Minister of Environment reacted cautiously to the idea, listing a whole series of difficulties, he did not go so far as to reject the compromise. Indeed, the proposal was repeated in its entirety in the final declaration of the conference, together with a request to the IRC to study the technical, financial, legal and ecological implications of the suggestion and its likely effect on the quality of drinking water supplies. The IRC was requested to report back to the ministers on these points and to submit a draft amendment to the convention.

It was to take more than a year for agreement to be reached. The final dispute concerned the measures required to reduce the salinity of Lake IJssel. These pumping works were estimated to cost FF 100 million ($16 million) and, as part of the total package, they would also be financed jointly by the five Rhine states. Germany originally objected to the idea of paying for measures in the Netherlands, but eventually gave up its resistance when it became clear that this was the last remaining obstacle to agreement (the Reinwater Foundation openly speculated on the possibility that the Netherlands might have suggested to Germany that it would take a rather more flexible stance on discharges of toxic substances into the Rhine in order to achieve agreement). The text of the new protocol to the convention was finally agreed on 11 April 1991. Ratification by the national parliaments was only necessary to approve the measures to be taken in the Netherlands, and was not expected to cause undue problems; the remaining elements of the agreement were adopted by the simple device of a joint declaration by the five countries and through internal agreements within the IRC.

TOWARDS 1998

It is likely to be 1993 before any further reductions in the discharges are made because of the time required by the parliaments of the five Rhine states to approve ratification of the protocol and for MDPA to expand its storage facilities. That will leave just five years before the Rhine Chlorides Convention expires in 1998. From then on, the salt stored near the mines will be gradually dumped into the river. It is a scenario which bears little resemblance to the objectives of the original 1976 convention.

The history of the court actions against the discharges provides little evidence to suggest that judicial procedures will prove any more effective in reducing the discharges. Three of the actions remained to be decided in early 1991, including that concerning the permit granted on 5 September 1985: the decision of the Administrative Court in Strasbourg on 3 August 1989 to annul the permit was contested by MDPA at the Council of State and, following his usual practice, the prefect of Haut-Rhin promptly granted an almost identical permit. None of the three final judgments which had been made had actually led to a reduction in the discharges. It is in any case likely to take several more years before all the appeal procedures open to the Dutch litigants and MDPA are exhausted. And by then it might be too late. In 1998 the first of the sylvite mines is due to close; by 2004 the economically recoverable reserves may be exhausted. All that the lawyers will then be able to fight over is the destination of the 13 million tonnes of salt which MDPA will bequeath to the Alsatian landscape.

CHAPTER FOUR

ACID DROPS

The European Community's acid emissions control policy

The European Community (EC) is probably the most criticized institution in the modern world. If its critics are to be believed, the EC's bloated bureaucracy works at a pace falling somewhat short of that of a snail; its policy-makers, cocooned from the unpleasantness of the real world in their luxurious Brussels offices, are incapable of devising a proposal even remotely relevant to the continent's burgeoning economic, social and environmental problems, while politicians of all allegiances and nationalities, ever faithful to the deep-rooted mistrust between Western Europe's patchwork of states, look upon the very idea of consensus on any pressing issue as they would a particularly nasty disease.

A more objective analyst of EC policy-making would, however, emphasize rather different aspects of the decision-making process, particularly that the special difficulties which the Community faces stem to a large extent from its peculiar political status. Although it is a supranational institution, with power to lay down binding legislation on ostensibly sovereign member states – an arrangement which is unique in the world – its decisions are made jointly by the 12 governments. It therefore falls short of a true federal system where a superior central government is charged with the power to act in matters which affect the federation as a whole. The result is a decision-making culture which has difficulty in transcending national interests. Constitutionally, substantial majorities are necessary for legislation to be adopted. More importantly, the political reality of nurturing cooperation between countries with such diverse conditions and cultures and with such long histories of mutual antagonism is that the Community cannot afford to alienate individual member states from its policies or to create schisms between different regions. It is therefore a rare event for measures to be taken which are strongly opposed by more than a single member state; in other words, proposals which infer that a small number of countries will bear a disproportionate share of the costs of implementing the measures will almost invariably be obstructed by the affected member states.

In the environmental field, an issue which tested the ability of the Community to take effective action in the face of a serious environmental threat was the attempt to develop an EC acid emissions control policy. The proposal at issue was the so-called Large Combustion Plant Directive. Conceived as the principal component of the EC's response to the problem of acidification, the goal was to substantially reduce the polluting emissions from coal-, oil- and gas-fired plants. Never before had a single proposal carried with it such high control costs. But while this particular case represents, in some ways, an extreme example of EC policy-making, it is a classic illustration of the difficulties encountered in resolving the equity dilemma: who benefits from the activities which cause environmental impacts and who suffers the damage? The corollary, particularly apposite in this case, is just as difficult to resolve: who pays for the costs of abatement measures and who benefits from the environmental improvements?

EC POLICY-MAKING

Now comprising 12 member states, the EC is administered

by four separate bodies:

- the European Commission – the initiator of policy proposals and the executive arm of the administration;
- the European Parliament – the democratic assembly of 518 directly elected members which acts as watchdog and adviser with limited powers to amend legislative proposals;
- the Council of Ministers – the decision-making body of the Community, comprising, for each policy proposal, the appropriate 12 ministers from the governments of the member states (such as the Ministers of Environment for environmental proposals);
- the Court of Justice – responsible for ensuring observance with the three treaties that established the EC.

By the usual standards of democratic nation states, EC policy-making is a curious process. Its logic, however, lies not in any pretension to direct democratic accountability, but in providing a vehicle to facilitate the sensitive negotiations which are necessary to reconcile the differences between a variety of national interests, political institutions, economies and cultures. For example, only the European Commission may introduce a formal policy proposal, giving it a far more prominent political role than national executives enjoy; and as will later become clear, this power of a non-elected body to define the policy agenda has far-reaching implications for the EC policy-making process.

One of the less appreciated achievements of the EC is the development of a Community environmental policy. From being almost exclusively the prerogative of national governments, the context of environmental policy has gradually, but radically changed since 1972, the year when the member states formally recognized environment protection as a proper objective of EC policy. Indeed, it has developed to the point where the main lines of national environmental policy are, to a large extent, prescribed by the EC, and there now exists a remarkably comprehensive body of EC environmental policy and legislation. This comprises a series of strategic Environment Action Programmes and about 200 items of environmental legislation, covering virtually all aspects of environmental protection – air pollution, water pollution, wastes, noise nuisance, hazardous chemicals, nature conservation and environmental assessment.

THE BACKGROUND TO THE PROPOSAL

Although the first serious concern over a possible link between air pollution and acidification arose in Sweden in the 1960s (and, indeed, was a pivotal factor in Sweden's decision to convene the Stockholm Conference on the Human Environment in 1972), it was not until the late 1970s that the problem came to be widely recognized as a major international environmental issue.[1] Formal recognition came with the signing by 35 countries of the 1979 UNECE Convention on Long-Range Transboundary Pollution (generally known as the Geneva Convention), although the text of the convention failed to meet the demands of Sweden and Norway that there should be an obligation to reduce sulphur dioxide emissions by up to 50 per cent. (Ironically, as will later become clear, one of the strongest opponents to this proposal was the Federal Republic of Germany, which was not even prepared to accept a standstill clause on sulphur dioxide emissions.)

The evidence linking air pollution and acidification continued to accumulate, and in 1982 a political event took

[1]It is not the purpose of this chapter to discuss the scientific debate on acidification and the likely causes of the problem. The essential fact as far as this analysis is concerned is the now widespread political acceptance of the hypothesis that air pollutants – principally emissions of sulphur dioxide, nitrogen oxides and ammonia – play a key role in the process of acidification.

place which was to prove crucial to the international regulation of acid emissions. At the Conference on the Acidification of the Environment, held in Stockholm on 28–30 June, the Federal Republic of Germany suddenly shifted from its previous public stance of doubting the seriousness of the acidification problem when the Minister of the Interior, Gerhart Baum, called on all states to combat polluting emissions, committing the Federal Republic to halve its own emissions of sulphur dioxide by 1985. Underlying this decision was the new evidence of serious damage to Germany's extensive forests, and an awareness of the growing political force represented by the Greens. In these circumstances, an aggressive campaign against acid emissions became a political imperative for the federal government, and since that moment Germany has been the most prominent political proponent of substantial emission reductions within the EC.

The first mark of the new German stance came with the introduction of draft legislation, the *Grossfeuerungsanlagen-Verordnung* or GFAV, aimed at limiting emissions of sulphur dioxide, nitrogen oxides, particulates, carbon monoxide and chlorine from large coal- and oil-fired combustion installations. In the context of air pollution legislation then prevailing in other EC countries, the new proposals represented a substantial tightening of emission control, although existing plants were to have up to 10 years in which to comply with the standards. It is nevertheless indicative of the strength of popular opinion then prevailing in the country that the proposals were further tightened during the passage of the bill through Parliament and that the regulations came into effect on 1 July 1983, less than ten months after their publication. Further, the measures were retained, virtually unchanged, by the more conservative Christian-Democrat/Liberal coalition which succeeded the Social-Democrat/Liberal government a few weeks after the proposals were announced.

Far-reaching as these measures were, they could have

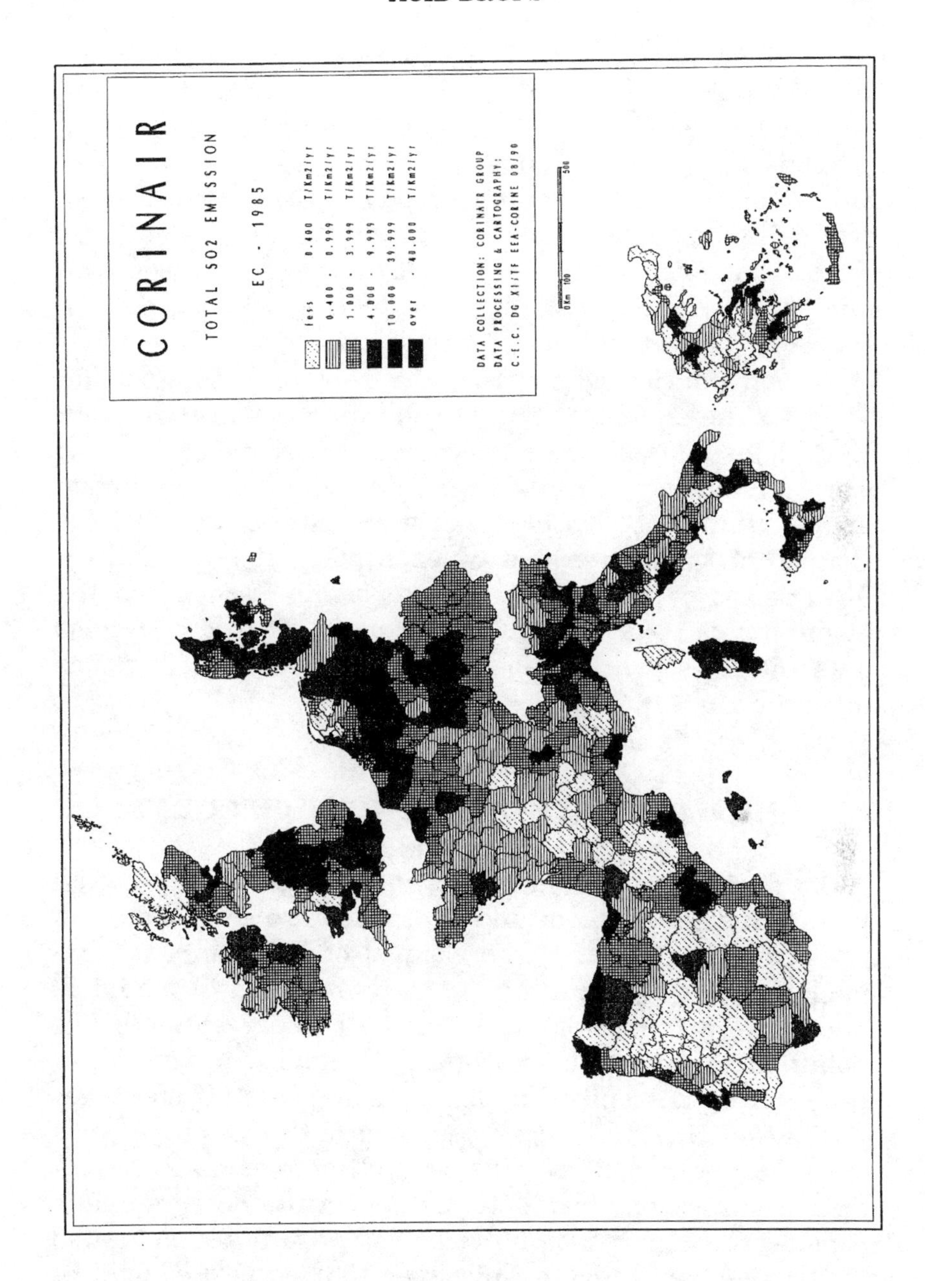

Sulphur dioxide emissions in the EC

only a limited effect on the German environment for the simple reason that over 50 per cent of the net acid deposition in the Federal Republic was imported from surrounding countries. The primary aim of the government was, therefore, to secure effective international control measures on acid emissions. Consequently, parallel to its own national programme of action, the federal government began an intensive lobbying campaign within the EC. It was considerably helped in this effort by a fortuitous alignment of responsibilities within the European Commission: not only was the Commissioner for the Environment a German, Karl-Heinz Narjes, with a German-oriented cabinet, so too were certain key officials with responsibilities relating to air pollution control policy. And with the presidency of the Council of Ministers, held by each member state in turn for a period of six months, due to pass to Germany at the beginning of 1983, the time was ripe to get acid emissions onto the EC's political agenda in the hope that decisions might be made quickly.

THE EMERGENCE OF THE DRAFT DIRECTIVE

A few days before he made his announcement at the Stockholm conference, Baum had submitted a memorandum to the periodic meeting of the Council of Environment Ministers requesting that a high priority be given to the introduction of a framework directive on the prevention of air pollution and that the European Commission draw up a proposal by the end of that year. Despite staff shortages at the Commission, Narjes was keen to have a proposal on the table as soon as possible and, despite some objections from Commission officials that the measure deserved more considered preparation, a preliminary draft was circulated to the member states in November 1982 – just two months after the German national proposals were first announced. The Commission's proposal was for two directives: a so-called "parent directive", laying down the broad framework for the control of emissions from industrial plants; and a "daughter directive", specifically aimed at acid emissions. There was, not surprisingly, a high degree of correspondence between the German and the EC proposals; as proposed, the daughter directive was to fix limit values for emissions of sulphur dioxide, nitrogen oxides and particulates from all large combustion plants with a thermal capacity exceeding 100 MW, the size of a large hospital boiler, in all member states. The limits suggested were 400 mg/m^3 for sulphur dioxide, 800 mg/m^3 for nitrogen oxides and 50 mg/m^3 for particulates, though a number of derogations were provided for to allow laxer standards to be adopted where abatement costs would otherwise be unreasonably high.

Although the proposal was discussed at the December meeting of the Council of Ministers, progress was at first slow, primarily because the meeting of national experts, organized to exchange opinions on the proposal, decided to give first priority to the framework directive. The German government nevertheless kept up the pressure, making good use of its EC presidency at the Summit of Heads of State in March 1983. This proved to be a major political success for the Federal Republic, for it secured not only an agreement that the widespread damage to forests required urgent and effective action, but also an undertaking that the Council of Ministers would give "rapid and urgent attention" to the air pollution control proposals being prepared by the Commission.

Their first chance to do so came within two weeks, when the proposal for a framework directive on the control of air pollution from industrial plants was formally submitted. The key provision as far as acid emissions were concerned was the article authorizing the Council, acting on a proposal from the Commission, to fix limit values for emissions from industrial plants, especially for eight groups of pollutants listed in an annex (which included sulphur dioxide, nitrogen oxides and other nitrogen compounds).

It was proposed that these limits should be agreed by a qualified majority of the Council.

With the draft framework directive now before the Council, the way was open for the Commission to continue the preparatory work on its proposal for a daughter directive on acid emissions. Despite the persistence of serious doubts in some quarters – particularly in the UK – that air pollution actually was a primary cause of acidification, the Commission made it clear, in replying to a question from the European Parliament, that it did not share these misgivings. Its determination to proceed with the proposal for a directive was clear, and this was reinforced by its commissioning of a cost-benefit analysis on reducing acid emissions, although the considerable scientific uncertainties allowed only a broad indication to be made of the total damage costs of acid pollutants in the EC and Scandinavia, with a probable total in the range $0.5–3.5 billion a year.

Pressure for firm action was also beginning to mount from outside the EC. Unhappy at their failure to secure European-wide sulphur dioxide emission reductions under the Geneva Convention, the Scandinavian countries took advantage of the first meeting of the Executive Board of the convention in Geneva on 7–10 June to table a proposal for a uniform reduction of sulphur dioxide emissions of 30 per cent by the signatory states between 1983 and 1993. Support for the idea came from Canada, the Federal Republic of Germany, Switzerland and Austria, the latter three countries also proposing corresponding reductions in emissions of nitrogen oxides. Both proposals were rejected, however, the major opposition coming from the US, the UK, France and the Eastern European states. But if no concrete action was agreed, the compromise final declaration noted "a recognition of the need to decrease effectively the total annual emissions of sulphur compounds, or their transboundary fluxes, by 1993/95 using 1980 emission levels as the basis for calculation". Moreover, although the first steps towards a European acid emission control policy had only just been taken, it was those three key elements – uniform emission reductions, a 1980 baseline and a 1993/95 deadline – which, for better or for worse, effectively set the policy agenda for the years to come.

The main elements of the Commission's acidification control strategy were first set out in a document published in November 1983. An array of measures were envisaged, ranging from encouraging member states to adopt stricter standards than the somewhat lax levels laid down in a 1980 directive on air quality standards for sulphur dioxide and particulates, to a new package of vehicle emission control measures. But the crucial proposal remained the daughter directive on acid emissions from large combustion plants. Indeed, it was clear that, were the directive to be approved in the form then being canvassed, the rest of the Commission's broad strategy would be largely superfluous. For by now a fundamental change had been made to the preliminary draft: new plants would be required to meet strict emission standards as originally proposed and, in addition, total emissions from all large combustion plants would have to be reduced in each member state by a uniform percentage. The reasoning behind this change, bearing in mind that the range of polluting sources varied from country to country, was to allow each member state to decide for itself how it might best reduce existing emissions most efficiently. Alas, not all member states were to be equally enthusiastic over the consequences of this flexibility.

The exact thinking of the Commission became clear on 15 December when the proposal for a draft directive was published. As expected, the formal proposal fell into two parts: emission standards for large combustion plants built (or substantially altered) after 1 January 1985, and obligatory reductions of total emissions from all installations. Emissions from new plants (now defined as those with a thermal capacity exceeding 50 MW) were to meet the limit values set out in Table 1. With regard to the total emissions from large combustion plants, each member state was to be required by the end of 1986 to draw up a programme for reducing emissions of the three pollutants. These reductions, to be achieved by 31 December 1995, were to be a minimum of 60 per cent for sulphur dioxide and 40 per cent for both nitrogen oxides and particulates, based on the level of emissions recorded in 1980. Finally,

Table 1: Proposed EC emission standards for new large combustion plants, December 1983.

FUEL	EMISSIONS				
	Sulphur dioxide		Nitrogen oxides		Particulates
	From 1.1.85 mg/m^3	From 1.1.96 mg/m^3	From 1.1.85 mg/m^3	From 1.1.96 mg/m^3	mg/m^3
Solid	400	250	800	400	50
Liquid	400	250	450	220	50
Gas	35	35	350	180	5

Less stringent limit values were applicable in certain circumstances, such as for solid-or liquid-fuelled plants with a thermal capacity below 300 MW, and the deadlines were extended by five years for plants below 100 MW thermal. The table has been simplified for presentational purposes.

in order to allow for differing circumstances in the member states, temporary derogations were to be permitted from the main requirements of the directive. These were to be granted at the discretion of Commission, and specifically in two cases:

- where the reliance on fuel with a high sulphur content was such as to lead to "disproportionate" abatement costs;
- where total annual emissions in a member state were so much lower than the levels in other member states that it would be difficult to reduce these by the amount required.

Pollution abatement on this scale had never before been

proposed by the Commission. All the more reason, therefore, to provide a thorough assessment of the likely costs of the measures, the projected environmental benefits, the distribution of the costs and benefits, and a carefully reasoned justification of the decision to opt for uniform national reductions in emissions from all large combustion plants, rather than adopting an approach which would take into account the varying situations in the different member states. Some discussion of the costs to be incurred by member states was included in the explanatory memorandum to the proposal, although this was confined to new plants only. The Commission bravely acknowledged that many of the factors likely to affect the costs of the measures were unknown, but taking a high assumption of 40 000 MW of new capacity at plants exceeding 300 MW by the end of the century, plus an unspecified contribution from smaller plants, an annual cost of approximately $1.15 billion was projected. Against this crude calculation was set the annual damage caused by sulphur dioxide and nitrogen oxides, taken from the consultant's report, roughly estimated now at a total of $1.4–4.1 billion a year. No detailed breakdown of the costs and benefits for each member state was included. Similarly, there was no discussion of the reasoning underlying the intention to require a uniform national emission reduction for all plants, or of the proposed cuts of 60 and 40 per cent, although a reference was made to the conclusions of two international conferences which advocated, according to the Commission, a reduction of at least 50 per cent in emissions of sulphur dioxide and nitrogen oxides if the soil's acid neutralization capacity was to be restored. In fact, the findings of the two meetings involved, at Stockholm in June 1982 and Karlsruhe in September 1983, were by no means so unequivocal and certainly did not extend to the question of whether such cuts should be applied uniformly across the 12 member states. In short, the draft directive was not only potentially the most expensive item of EC environmental legislation ever proposed, eventually to cost the member states some $20–30 billion, it was also likely to become the most controversial.

INTERLUDE: SOME THOUGHTS ON THE CHOICE OF A CONTROL STRATEGY

The decision to opt for a control strategy based on uniform emission standards for new plants and uniform national emission reductions for all plants in aggregate deserves further consideration. It could be argued that the application of EC-wide emission standards to newly built large combustion plants was a straightforward and politically defensible strategy, simply reflecting the "best available technology" control philosophy (although the actual limit values proposed and the extent to which the cost of abatement in each particular case should be taken into account in setting the limits remained matters for discussion). That is not to say, however, that it was the most logical or most appropriate approach. But the concept of uniform national emission reductions based on the 1980 emissions from large combustion plants was an altogether more controversial proposal. For what was being proposed was the establishment of 12 national "bubbles" for an arbitrarily chosen year (1980), the obligation to secure uniform percentage reductions in those emissions and the fixing of an apparently arbitrary deadline for the achievement of the reductions (1 January 1996). Quite apart from the fact that no member state made use of a binding national bubble to limit its atmospheric emissions, there were any number of alternative strategies which could have been adopted, many of which at least had the merit of reflecting rational and relevant control principles. It is worth pausing to consider the various alternative approaches which might have been adopted.

EC bubble

In the first place, a maximum ceiling on total EC emissions of acid pollutants could be agreed and then allocated as 12 "sub-bubbles" amongst the member states. In defining the bubble, a basic decision would have to be made as to the extent to which the environmental impact of EC acid emissions on non-EC countries should be taken into account. The bubble might then be calculated in a number of different ways:

- agreeing international minimum emission reductions, perhaps enhanced through the use of some political formula to take into account certain variables (such as national/EC income or industrial structure);
- agreeing a fixed quantity of acid emissions per head of population;
- agreeing a fixed quantity of acid emissions per hectare;
- making internationally equivalent emission abatement investments (such as $x per installed MW of combustion plant);
- making comparable emission abatement investments on the basis of national/EC income (the "comparable effort" principle);
- equalizing marginal abatement costs (perhaps combined with capital transfers);
- using some form of cost-benefit analysis;
- working back from a generalized (rather than locally specific) level of acid deposition or environmental acidification.

Given an EC bubble, many of the above principles might again be applied to allocate the reductions.

Reducing transboundary fluxes

In terms of environmental management, the major problem

of acid pollution is its transboundary nature: some countries are net importers of acid emissions, others are net exporters. It might therefore be argued that the object of the control process should be to reduce these transboundary fluxes. This might be done, first, at international level, to agree maximum fluxes across the EC's borders and, second, within the EC itself to establish comparable fluxes between the member states. Geographical factors such as the location of each member state, prevailing wind speed and direction, and the location of industrial concentrations would inevitably lead to wide variations in abatement costs between the 12 countries.

Technology standards

The simplest and most straightforward option would be to introduce EC-wide emission standards for all large combustion plants, both new and existing installations (as originally proposed by the Commission). Alternatively, a fixed level of emission abatement might be laid down for all plants, such as 90 per cent flue-gas desulphurization (FGD), thereby accounting for the problems caused by varying sulphur contents of indigenous coals. These are so-called technology standards, based either on the "best available technology" or the less strict "best practicable means", or indeed a combination of the two (such as best available technology for new plants and best practicable means for existing installations). Again, abatement costs would vary considerably from country to country depending on the degree to which fossil fuels contributed to national electricity generation. Moreover, with no direct relation to air quality, it might be argued that tougher action should be taken in highly industrialized, heavily polluted regions while laxer measures could be tolerated in remote areas.

Uniform air quality standards

Tackling acidification by setting a generalized uniform air quality has the clear advantage of basing control on environmental grounds. An homogeneous air quality standard would not, however, protect those regions (such as much of Scandinavia) where the surface environment has a low acid neutralization capacity unless an exceptionally rigorous standard were to be adopted (and it might then be argued that such a rigorous standard would be excessively strict for large parts of the EC). Local and regional air pollution control authorities would also have great difficulty in regulating acid emissions solely on the basis of an air quality standard since they would have neither sufficient information on the exact long-range impacts of the emissions from any particular plant within their jurisdiction, nor would they be able to exert any control over a large proportion of the acid pollutants transported into their districts.

Critical loads

The most sophisticated of the control principles which might be applied to acid emissions is the critical loads concept. Considerable research has been carried out in recent years into the sources of acid emissions, the long-range transport of pollutants in the atmosphere, acid deposition rates and the capacity of various environments to neutralize acidic inputs. Putting these data together opens up the possibility of determining the maximum tolerable input of acid deposition in any particular area and then, working backwards along the environmental pathway, setting limits on the emissions of acid pollutants upwind to ensure that this input is not exceeded. Here is a tool which, properly applied, provides a rigorous scientific basis for reducing and allocating reductions in acid emissions. The main problem is the huge volume of data which would have to be processed to determine the emission:deposition budget for any single locality. And it should not be forgotten that incorporating such a complex calculation into an international pollution control regime would open up an array of practical and legal difficulties. The key question is whether some simplified, manageable form of critical loads could be developed which could be formalized into an international air pollution agreement.

These, then, are the various control principles which could have been applied to the problem of acid emissions. As already described, the Commission elected to propose a joint system: uniform national emission reductions for all large combustion plants taken together and technology-based uniform emission standards for new installations. But because of the wide range of options which were available to the policy-makers and the different functions and distributions of their respective costs and benefits, the scope for disagreement as to the most appropriate principle of control was considerable. And having made its decision, the Commission managed to increase the potential for conflict by failing to provide any rigorous arguments in support of its case. If securing agreement on such a contentious issue was always likely to be a hard-fought battle, the proposal now on the negotiating table only provided all sides with new ammunition.

THE NEGOTIATIONS

Just as the preparations for the hard bargaining were being made in each of the member states, an event took place in Ottawa which was to change the whole nature of the international debate on acid emissions. Although their proposal for a uniform 30 per cent reduction in 1980 level emissions of sulphur dioxide by 1993 was rejected at the June 1983 meeting of the Executive Board of the Geneva

Convention, the Scandinavian countries doggedly persisted with their lobbying and managed to persuade a number of other countries to adopt the goal in a formal declaration. Thus it was that, in March 1984, representatives from 10 countries came together in Canada to sign their own separate agreement, each committing itself to reduce its 1980 emissions of sulphur dioxide by a minimum of 30 per cent by 1993. Some countries went even further, promising greater cuts, as shown in Table 2. They also agreed to take action to reduce emissions of nitrogen oxides by 1993 at the latest. The "30% Club" was born.

Table 2: The 30% Club, March 1984.

Country	Agreed cut in 1980 SO_2 emissions	Deadline
Austria	30%	1993
Canada	50%	1994
Denmark	40%	1995
Finland	30%	1993
France	50%	1990
Germany	50%	1993
Netherlands	40%	1995
Norway	50%	1994
Sweden	30%	1993
Switzerland	30%	1993

By now the ball was well and truly rolling. Further impetus was added at the Multilateral Conference on the Environment, held in Munich on 24–27 June, where environment ministers from the UNECE states requested the Executive Board of the Geneva Convention to prepare a proposal for reductions in sulphur dioxide emissions (or their transboundary fluxes) by 1993 and to extend its work to the control of nitrogen oxides. With so much activity now going on, the time was clearly ripe for decisive action by the EC.

The opening shots in the debate were fired from the European Parliament. On 16 November 1984, 11 months after the draft Large Combustion Plant Directive was published, the Commission's proposals were rejected in a resolution as "totally inadequate". The Parliament's main argument was that the 1980 emissions of all three pollutants from existing plants should be reduced by 50 per cent by 1990 and by 75 percent by the end of 1995. At the same time, existing plants should also have to comply by 1 January 1991 with an exceptionally intricate system of strict emission standards. These limits would vary according to the capacity of the plant, the expected number of remaining full-load hours and the type of fuel. For solid-fuel plants with a thermal capacity exceeding 300 MW and with a life expectancy exceeding 30 000 full-load hours a limit value of 400 mg/m^3 was proposed, equal to the emission standard in the draft directive for new plants. This proposal reflected the approach adopted in the German GFAV. For new plants, the Parliament wanted the second-stage emission standards to apply from the end of 1990 rather than from 1 January 1996; and instead of a dual system of limits for plants above and below 300 MW, it proposed making a further distinction for plants with a capacity below 100 MW. Finally, in order to ensure that the limited supplies of low-sulphur fuel were used primarily in smaller plants where FGD was more expensive, it recommended that new plants of 300 MW or more should be required to remove at least 85 per cent of their sulphur dioxide emissions, and plants in the 100–300 MW range at least 60 per cent. With the Council of Environment Ministers due to discuss the draft directive at its meeting on 6 December, the question was whether the EC's decision-making process could keep up with the fast-moving international debate on acid emissions.

Although included on the agenda for the Council meeting, a serious obstacle to detailed discussions of the proposal had arisen in the form of a parallel draft directive from the Commission on vehicle emissions. Also designed to reduce acid emissions, this proposal proved to be so controversial that it absorbed most of the negotiating time at three successive Council meetings until an outline agreement received majority support in June 1985. It nevertheless became clear at the December meeting that the member states could be divided into three broad groups in their response to the large combustion plant proposals: strong support from Germany and the Netherlands, qualified support from France, Denmark and Belgium, and strong opposition from the UK, Greece, Ireland and Luxembourg. Italy refrained from taking a definite position. (At this time Spain and Portugal had yet to accede to the Community.) Greece, Ireland and Luxembourg, arguing that their emissions made a negligible contribution to acidification, pressed for partial exemption from the proposed restrictions. The UK, as the largest emitter of sulphur dioxide in Western Europe, was in a different position altogether. Its argument was that the scientific case linking air pollution with the environmental damage attributed to acidification was far from proven, and in the absence of such unequivocal evidence it would be imprudent to make the huge investments necessary to achieve the proposed emission reductions.

Table 3: Revised EC emission standards for new large combustion plants, March 1985.

FUEL	PLANT CAPACITY	EMISSIONS				
		Sulphur dioxide		Nitrogen oxides		Particulates
	MWth	From 1.1.85 mg/m^3	From 1.1.96 mg/m^3	From 1.1.85 mg/m^3	From 1.1.96 mg/m^3	mg/m^3
Solid	›300	400	250	650	200	50
	300-100	1200	1200	800	200	50
	‹100	2000	2000	800	400	50
Liquid	›300	400	250	450	150	50
	300-100	1700	1700	450	150	50
	‹100	1700	1700	450	150	50
Gas	All	35	35	350	100	5

The table has been simplified for presentational purposes.

By now, however, the target was on the move yet again, for the Commission was already working on a revised proposal. Published on 22 March 1985, the amendments were largely prompted by the opinion of the European Parliament. The main changes concerned the proposed emission standards for new plant, as shown in Table 3. First, in response to the opinion of the European Parliament, a three-tier hierarchy of plants was introduced and, second, the limit values for emissions of nitrogen oxides were tightened by a significant margin. The changes nevertheless fell far short of the Parliament's demands and were not well received in the assembly.

Meanwhile, the 30% Club had caught on in a big way, and by April 1985 the number of members had increased to 21, including a sizeable East European contingent.[2] Moreover, the Executive Body of the Geneva Convention had also drawn up a draft protocol to the original convention which formalized a 30 per cent reduction in 1980 level emissions of sulphur dioxide (or their transboundary fluxes) by 1993, and at a meeting in Helsinki in July the members of the 30% Club duly signed the document, now known as the Helsinki Protocol. Of the ten EC member states, only three refused to sign the protocol: the UK, Greece and Ireland. Again it was the UK which was the most notable dissident, arguing that although its total emissions of sulphur dioxide had declined by 25 per cent since 1980, it could not guarantee a further reduction of 5 per cent by the 1993 deadline.

An interesting possibility raised at this time was whether the EC itself should become a member of the 30% Club. This would have involved a commitment by the Council of Ministers that the aggregated sulphur dioxide emissions from all 10 member states should be reduced by 30 per cent by 1993.

---

[2]The 11 new members were Belgium, Bulgaria, the Byelorussian Soviet Socialist Republic, Czechoslovakia, the German Democratic Republic, Hungary, Italy, Liechtenstein, Luxembourg, the Ukrainian Soviet Socialist Republic and the Soviet Union.

The crucial flaw, as some of the countries most supportive of substantial reductions soon spotted, was that with some member states already committed to greater emission cuts than 30 per cent, other countries could take less drastic measures and still ensure that the EC target as whole was achieved. The European Commission soon dropped the idea.

Because of the vehicle emissions dispute, it was not until the December meeting of the Council of Ministers that further serious discussion of the Commission's proposals was possible. Even then, little progress was made. An attempt was made by the Commission to isolate the UK (and to subject it to decisive political pressure) by working on a statement of principles which might prove acceptable to the other nine countries. This was to include certain exemptions for the smaller countries, because of their minimal contribution to acidification, and concessions to Italy which, with plans to expand its coal-fired electricity generating capacity, was now firmly opposed to the proposal. But the attempt failed to persuade Italy to shift its ground, and further discussion was deferred to the next meeting.

GOING DUTCH

A new factor was about to complicate the proceedings, for on 1 January 1986 two countries were due to accede to the EC: Spain and Portugal. Industrial development in both countries was less advanced than in the EC as a whole, with the result that a strict acid emissions control measure would act as a serious impediment in their attempts to expand a relatively small generating capacity. Early agreement on the proposal seemed further away than ever. Faced with this unhappy prospect, the new president of the Council of Ministers, the Dutch Minister of Environment, Pieter Winsemius, decided to develop a new negotiating framework which departed from the original proposal in several important ways.

The groundwork for this initiative had, in fact, already been made in 1985 during preparations by the Netherlands for its six-month presidency. By then, the objections of certain countries to the proposal were clear, and the problems likely to arise after the accession of Spain and Portugal had already been anticipated. The idea launched by policy-makers inside the Dutch Ministry of Environment was that a revised proposal should be formulated, which took account to some degree of the acid pollution caused by each member state, its level of industrial development and its emission trends. To this end a confidential assessment was commissioned from specialist consultants to examine the extent to which various member states would reduce their acid emissions in the absence of a directive. The outcome was a set of explicit principles by which the reductions could be determined. These principles included: the fixing of a bubble for the EC as a whole; the application of the "best available technology not involving excessive cost" standard; a two-stage reduction in emissions from all combustion plants (which might allow substantial reductions to be deferred to the second stage by a few member states); and a number of variables to account for differing national circumstances – the level of national emissions, the relative contribution to pollution in Europe, national economic conditions and the type of fuels available. The result was a proposal which, for the first time, explicitly moved away from the concept of uniform emission reductions. This was a major advance both politically and conceptually and was generally welcomed by the other member states. Indeed, 11 of the 12 ministers approved the new principles – though once again it was the UK which remained outside the fold.

Given this broad agreement, it was decided to elaborate the new principles into specific emission control proposals in time for the next Council meeting in June, and by May a draft paper had been circulated to the member states. This concentrated on the sulphur dioxide problem, proposing a two-stage reduction in total EC emissions from large combustion plants of 45 per cent by 1995 and 60 per cent by 2005 on a 1980 level baseline. The proposed allocation of the 45 per cent reduction was calculated "objectively" for each member state on the basis of its total combustion plant emissions, its contribution to acid deposition in other member states, its installed generating capacity, its gross domestic product per head of population and its "need for economic development and for large scale use of 'difficult' indigenous fuels". The resulting national reductions for the eight largest countries fell between 27 per cent and 63 per cent, as shown in Table 4. The remaining countries, Greece, Ireland, Portugal and Luxembourg, would be exempted from cuts because of their limited emissions and minimal contribution to acidification. This "objective" reduction was then adjusted to some extent to take into account planned or expected reductions in each member state. Finally, calculations of the likely reductions incorporating the substantial cuts projected for some countries were also included.

Table 4: Proposed reductions in sulphur dioxide emissions from large combustion plants, May 1986.

MEMBER STATE	PERCENTAGE REDUCTIONS IN 1980 SO_2 EMISSIONS BY 1995			
	1 Planned without EC action	2 "Objective" contribution to 45% total	3 Proposed minimum reduction	4 Likely reduction combining 1 and 2
Belgium	50	41	50	50
Denmark	50	33	50	50
France	60	57	50	60
Germany	70	63	50	70
Italy	30	27	40	40
Netherlands	60	36	50	60
Spain	10	32	10	10
UK	20	59	40	40
EC total	30	45	45	45

No calculations were circulated for allocating the cut of 60 per cent by 2005, although it was suggested that the Commission should draw up a proposal by 1990. It was also suggested that the first-stage emission standards for new plants should remain largely unchanged but that the Commission should issue revised limit values by 1990 for the second-stage limits, to take effect from 1996 and to be based on the "best available technology not involving excessive costs". Finally, as far as emissions of nitrogen oxides were concerned, a two-stage approach based on the same control principles as applied to existing sulphur dioxide emissions was proposed, with an ultimate reduction in 1980 emissions of 40 per cent, to be achieved in 2005 by way of an unspecified intermediate cut by 1995.

The Netherlands was aware before the Council meeting that Germany was unhappy about the idea of a "three-speed" Community (a reflection of the 50:40:10/0 percentage cuts) and also that the Commission itself saw the proposals as a significant weakening of its own draft which aimed for a 60 per cent reduction by 1995. But an hour before the start of the meeting the Dutch delegation had been given to understand that, although the Environment Commissioner, Stanley Clinton Davis, would openly criticize the new approach, he was nevertheless willing to concede if an agreement seemed possible. During the meeting, Germany duly objected to the three-speed approach and the UK reiterated its general objection to the draft directive. The Dutch response was to suggest a uniform first-stage reduction in the 1980 sulphur dioxide emissions of 40 per cent with the possibility of Spain and the smaller countries being granted derogations to account for their special positions. Then Germany, backed by the Commission, proposed that this relaxation should be compensated for by bringing forward the second-stage reduction of 60 per cent from 2005 to 1998. Clinton Davis also made clear, in a half-hour contribution, that he was, in fact, firmly opposed to the new compromise, finding it far too lax, particularly regarding the contribution required from the major emitting countries, and arguing that as a result the EC as a whole might not match the Helsinki Protocol goal of a 30 per cent reduction from all sources by 1993.

This failure to reach an agreement was the wrong result at the wrong time. Not only did the new proposal offer the prospect of a generally acceptable political compromise, but broad agreement at this stage held the promise of a substantial total reduction in acid emissions. It would also have exerted powerful political pressure on the UK to fall into line with its EC partners and – perhaps cynically – it might have been approved by Spain and Portugal while they were still finding their feet in the Community (although Spain had already made clear that it was opposed to EC-wide emission reductions). With hindsight, there is no doubt that the strong opposition from the Commission and Germany was a tactical error, for the agreement which was eventually reached after further protracted negotiations was to require less stringent measures than the Dutch proposal of May 1986.

TAKING TURNS

With negotiations moving into the second half of 1986, a further complication arose with the passing of the presidency of the Council of Ministers to the UK, the most fervent opponent of the draft directive. Even in the UK, however, things were changing. On 11 September the Secretary of State for the Environment, Nicholas Ridley, announced that the government had provisionally authorized the Central Electricity Generating Board (CEGB) to retrofit up to 6000 MW of existing generating capacity with FGD equipment between 1988 and 1997. A final decision, however, would be taken only after the joint Royal Society/Scandinavian Academies of Science report on surface-water acidification, due in spring 1987, had been studied. Moreover, it was confirmed that all new coal-fired power stations would henceforth be fitted with FGD. Although this decision amounted to the de facto recognition by the UK government of a link between emissions of sulphur dioxide and nitrogen oxides and environmental acidification, it was nevertheless far from clear what the precise effects of the measures would be. Fitting FGD to three 2000 MW power stations would remove at the very most 405 000 tonnes of sulphur dioxide a year, about 15 per cent of the total emissions from power stations. But this would be less if the stations concerned burned coal with a sulphur content below 2 per cent or if they were downgraded in the CEGB's merit order because of the 1 per cent drop in combustion efficiency caused by FGD. And with emissions from UK large combustion plants projected to creep back up to their 1980 level by the late 1990s, the programme, even under theoretically ideal conditions, would fall well short of the proposed first-stage cut of 45 per cent.

The approach adopted by the new British president of the Council of Ministers, Minister of Environment William Waldegrave, on the crucial issue of sulphur dioxide emissions was to develop the principles formulated in the Dutch compromise, but to incorporate three new elements into the calculations. First, reductions should be based on total national emissions, not just those from large combustion plants. Second, these emissions would be reduced in three stages (1995, 2005 and an unspecified later date) to produce total EC reductions of 30 per cent, 45 per cent and 60 per cent. Finally, the second- and third-stage cuts would be related to per capita sulphur dioxide emissions: 50 kg maximum per person by 2005 and 30 kg in the third-stage. On the basis of these criteria, the proposed reductions were allocated as set out in Table 5. In addition, Waldegrave proposed that detailed negotiations on the first-stage (1995) emission standards for new plant should commence immediately to prepare the way for the Commission to make a further proposal for tighter second-stage (2005) limits by 1990. He suggested, however, that these standards should vary according to the capacity of the plant: laxer standards for smaller plants, tighter standards for larger plants. For emissions of nitrogen oxides, Waldegrave further proposed that the first-stage cuts should be based on combustion modifications – low-NO_x burners – and that the Commission should draw up proposals for second-stage reductions by 1990, taking into account any advances in nitrogen oxide abatement technology.

Table 5: Proposed reductions in total sulphur dioxide emissions, November 1986.

Member state	Percentage reductions in total 1980 SO_2 emissions 1995	2005	Later
Belgium	50	50	
Denmark	50	50	
France	50	50	
Germany	50	50	Still
Greece	0	0	to
Ireland	0	19	be
Italy	30	40	allocated
Luxembourg	0	0	
Netherlands	50	50	
Portugal	+25	+25	
Spain	5	28	
UK	21	40	
EC total	30	45	60

The outcome of the negotiations on 24 November indicated that agreement was even further away than five months before. No member state, with the exception of the UK, was enthusiastic about the proposal. Spain and Ireland emphasized the difficulties which they faced with the 1980 baseline as a result of their expanding electricity generating programmes, the Dutch pointed out that their principle of "comparable effort" was given insufficient weight, and Germany and the Commission objected to the distant deadlines, particularly that for the third-stage reductions. It was also noted that the proposal opened up the way for the UK to escape any stringent obligations: not only had the fall in its total emissions of sulphur dioxide since 1980 already passed its target for 1995, but its emissions for 2005 after the three-station FGD programme and an expanded nuclear programme were likely to fall by the required 40 per cent. Moreover, with the likelihood that other member states would exceed their first-stage targets and thereby bring the 1995 EC aggregate reduction beyond 30 per cent, there was the possibility of a subsequent demand to relax the second- and third-stage cuts for certain countries on the grounds that less need be done to meet the 45 and 60 per cent targets.

This new failure left the initiative with Belgium, which took over the presidency of the Council of Ministers on 1 January 1987. Through an informal meeting of the environment ministers in February, Belgium proceeded to further elaborate the Dutch proposal, discarding much of the UK effort in the process. Returning to the original concept of dealing only with emissions from large combustion plants and the Dutch idea of a two-stage reduction, it introduced the notion of "emission credits" to take into account both the reductions achieved by certain countries before the 1980 baseline and the expanded generating programmes of the less advanced member states which had substantially increased their emissions since that date. No precise figure was suggested for the pre-1980 credits, but to allow for generating capacity installed between 1980 and the date on which the directive was eventually agreed, the proposed first-stage reduction in 1980 level sulphur dioxide emissions of 40 per cent by 1993 was to be reduced by 40 per cent of the "new" emissions (equivalent in aggregate to 40 per cent of the emissions on the date the directive was agreed for those countries which had increased their emissions since 1980). Discussions had also taken place on the emission standards for new large combustion plants, but the prospects were not encouraging. Most member states were satisfied with the lower capacity limit proposed by the Commission of 50 MW, but because of the relatively higher costs required to abate emissions from smaller plants the UK preferred a threshold of 100 MW and Ireland 150 MW. The capacity threshold at which full FGD should be required was also disputed: Germany and the Netherlands supported the proposal for 300 MW, the UK pressed for 700 MW. Further, the UK and Spain were unhappy about applying uniform limit values to emissions from plants fitted with FGD since those burning high-sulphur coal would require more sophisticated abatement equipment.

It therefore came as no surprise when the Council meeting on 19–20 March again failed to agree on the directive. Indeed, by now the proposals were becoming so complex that just about every member state was working to a different set of emission data and reduction targets. The Commission had also calculated that the emission credits might act to prevent the proposed EC reduction targets of 40 per cent in 1993 and 60 per cent in 1998 being achieved, and some countries still felt that the UK was being treated leniently. A formal declaration was nevertheless forced out of the ministers to the effect that emission reductions would reflect "comparable effort" and take account of "different conditions" in the member states. It was also notable that, although the UK had been granted an emission credit equal to the reduction of 500 000 tonnes in its large combustion plant emissions prior to 1980, its first- and second-stage reduction targets of 26 and 46 per cent respectively would still require a substantially larger FGD retrofitting programme than the three units already proposed – perhaps seven large units by 1998. The new plant proposals were also blocked now that the UK had been joined by Spain, Greece and Ireland in pressing for a higher threshold of 100MW for the application of emission standards to new plants.

Little progress was made at the Council meeting on 21 May when discussions continued on the Belgian proposals. By now Spain was asking for EC support in financing any major FGD programme for its power stations, and requesting that its first-stage target be deferred from 1993 to 1997 and, together with Italy, its second-stage reduction from 1998 to 2002. The UK also pressed for a deferral of its two targets, to 1995 (because its three-station FGD programme could not have the necessary effect any earlier) and 2005 (when its nuclear programme would have allowed older coal-fired plants to be decommissioned).

The second half of 1987 saw Denmark in the hot seat. Its own variation on the numbers game was to revert to the three-stage reduction philosophy and to juggle further with the cuts calculated from the Belgian proposal in order to provide extra leeway for those countries which had problems with the deadlines. The new allocations added up to total EC reductions in sulphur dioxide emissions of 34 per cent in 1993, 48 per cent in 1998 and 77 per cent in 2010, as shown in Table 6. Denmark was also willing to increase the threshold at which strict emission standards would apply to new plant from 50 to 100 MW, to compromise on 500 MW as the capacity at which full FGD should be required, and to offer member states the option of complying with a minimum desulphurization rate (from 30 per cent for small plants up to 90 per cent for the largest) to overcome the problems caused by applying a uniform emission standard to plants burning high-sulphur coal.

Although there was a cautiously favourable response during a meeting of national experts in October, little time was devoted to the proposal, probably due to Denmark's desire to agree a number of other items of draft legislation and because it had by then become clear that several member states had reservations about the desirability of committing the EC to emission reductions as far ahead as 2010. The result was that Germany – the initiator of EC action on acid emissions in 1982 and architect, as president of the Council of Ministers in 1983, of a joint declaration by the EC heads of state that the issue merited urgent attention – found itself five years later once more in the president's chair and with the proposal still on the table. Under these circumstances, the desire to secure agreement was a political imperative for Germany, and there were great expectations that its Environment Minister, Klaus Töpfer, would exert all his political influence to get a revised proposal approved.

Table 6: Proposed reductions in sulphur dioxide emissions from large combustion plants, October 1987.

Member state	Percentage reductions in 1980 SO_2 emissions 1993	1998	2010
Belgium	40	60	80
Denmark	40	60	80
France	40	60	80
Germany	40	60	80
Greece	40	40	60
Ireland	20	20	60
Italy	40	53	80
Luxembourg	40	50	60
Netherlands	40	60	80
Portugal	23	15	60
Spain	34	45	70
UK	22	33	80
EC total	34	48	77

THE DEAL

The main elements of the new variant became clear in February. Germany decided to retain a three-stage approach for reductions in sulphur dioxide emissions, this time with the combination for the EC as a whole of 40 per cent in 1993, 60 per cent in 1998 and 70 per cent in 2003. Emission credits for post-1980 increases in generating capacity were then built in (pre-1980 credits were no longer mentioned) and further downward adjustments made to take some account of special objections by certain countries, as shown in Table 7. However, after doing this, total EC reductions for the three phases came out at 25, 43 and 60 per cent. For emissions from new plants, Germany reverted to the original proposal for a minimum threshold of 50 MW, but with the limit values for plants in the 50–100 MW range to be agreed at a later date. The Danish compromise of 500 MW as the threshold at which full FGD would be required was retained, although the option of complying with a specific desulphurization rate instead of a uniform emission standard was restricted to a range of 60–90 per cent, depending on the plant's capacity. As a proponent of substantial reductions in emissions of nitrogen oxides, Germany also took the initiative to prepare a proposal for a two-stage uniform reduction of 25 per cent in 1993 and 40 per cent in 1998. These cuts would, however, be based on "adjusted" 1980 emissions by again taking post-1980 increases into account.

Table 7: Proposed reductions in sulphur dioxide emissions from large combustion plants, February 1988.

Member state	Percentage reductions in 1980 SO_2 emissions		
	1993	1998	2003
Belgium	40	60	70
Denmark	34	56	67
France	40	60	70
Germany	40	60	70
Greece	6	6	21
Ireland	+25	+25	+10
Italy	27	39	63
Luxembourg	40	50	60
Netherlands	40	60	70
Portugal	+29	+42	+26
Spain	0	24	37
UK	26	46	70
EC total	25	43	60

It quickly became clear at the Council meeting on 21 March that no progress was going to be made. It was the UK which came in for the most criticism, for although it was expected that the new British Environment Minister, Lord Caithness, would object to the new targets of 26, 46 and 70 per cent, claiming that it would be impossible to meet the first because new FGD equipment could not be installed in time and that up to nine FGD units would have to be retrofitted to meet the 1998 target, in the event no negotiations took place at all. Discussions ended when Lord Caithness refused to start negotiating until the 50 MW threshold for new plants was raised to 100 MW, on the grounds that this difference accounted for just 3 per cent of the sulphur emissions from large combustion plants but would involve unreasonably high abatement costs. Never in the five-year history of the proposal had tempers become so frayed.

As a result of this deadlock the whole issue threatened to get out of hand. The greatest obstacle to securing agreement was, of course, the UK, the largest emitter of sulphur dioxide in Western Europe. The British negotiators had persistently resisted any reduction target which would have involved a cut in 1980 sulphur dioxide emissions from large combustion plants of more than 15 per cent by the late 1990s. In particular, the 1980 baseline was seen as arbitrary (though the UK was by no means the member state most disadvantaged by this choice), but in reality the question revolved around the fact that the UK exported about two-thirds of its sulphur dioxide emissions while importing only about 12 per cent of its total sulphur deposition. In national economic terms, the costs of an extensive FGD programme were seen by the British government as outweighing the benefits by a substantial margin. But given the environmental impact of UK emissions on northwest Europe, a reduction of just 15 per cent would lead to only minimal improvements in the acidification levels elsewhere, even if the UK's easterly neighbours were to implement stringent control measures.

Of the other large member states, Spain was raising the most serious objections to the draft directive. The problem for Spain lay in its expanding generating programme and its indigenous high-sulphur coals which, it argued, were unduly penalized by the proposals. It was consequently looking for concessions which would allow less stringent abatement measures to be applied to these new plants (such as a maximum desulphurization rate of 60 per cent). The interesting issue here, as Spain well appreciated, was that its emissions made only a relatively minor contribution to the general problem of acidification, and any concessions could therefore be justified on environmental grounds. But it had proven to be beyond the creative skills of the negotiators to devise a formula which would have allowed Spain some latitude without doing the same for other countries, in particular the UK.

Italy was in a similar position to some extent, except that it was confident that it could achieve a reasonable overall reduction in its sulphur dioxide emissions by the mid-1990s (it was, after all, a member of the 30% Club). But a particular problem was the strict emission standards proposed for new plants, since much of Italy's indigenous coal has a high sulphur content. Italy would therefore have preferred to see optional limit values agreed for the first stage (up to 1993/95).

Finally, France, although ostensibly supporting strict emission controls, also objected to the proposed limit values for new plants. Because of the high level of nuclear base load in its electricity generating capacity, a large number of fossil-fuelled stand-by generators were used in order to cope with peak loads. But equipping new stand-by generators with FGD is relatively expensive given their small capacities and the limited use to which they are put. France was therefore pressing for special concessions to apply to such plants.

Other unresolved issues relating to new combustion plants

included: the minimum capacity at which emission standards were to apply (50 MW or 100 MW); the extent of any sliding scale for these limit values so as to allow laxer controls for the smaller plants; and the threshold at which full FGD was to be required (300 MW, 500 MW or 700 MW). There were, of course, any number of ways in which these variables might be combined. And as if that were not enough, the separate questions of nitrogen oxide and particulate emissions had hardly been broached. But the other side of the coin was that neither Germany nor the other member states which had persistently campaigned for strict acid emission control measures – France, the Netherlands, Denmark and Belgium – were prepared to accept an agreement with no teeth, since they had no interest in a directive which offered only a marginal reduction in environmental acidification.

Despite the gloomy prospects for progress, each of the main contestants had its own powerful reason for securing agreement. For Germany, agreement would represent a considerable political success given the prominent role that the acidification issue played in its domestic politics. The UK government, committed to privatizing its electricity industry, could not afford to allow the uncertainty over the scope of future emission abatement investments to continue for much longer as this would discourage potential investors. Both countries were also well aware that Germany had but one more chance as president of the Council of Ministers, at the meeting on 16 June, to come up with a new variation for substantial cuts in acid emissions. The presidency then passed to Greece and, in January 1989, to Spain, two countries which had not shown a great deal of enthusiasm for the proposal and which could not be expected to expend much effort on securing agreement.

The result was that in May, after some of the heat had dissipated from the conflict at the March meeting, informal discussions took place between the German and British ministers in order to examine the scope available for a compromise agreement. It was reported that the talks were constructive, though in the light of the events at the following Council meeting, the intriguing question is to what extent the course of the negotiations was in fact anticipated by Töpfer and Caithness. The reason for this curiosity is because the discussions came to be dominated not by the Large Combustion Plant Directive but by the issue of EC standards for polluting emissions from small cars – despite the fact that Klaus Töpfer's highest priority was the agreement on acid emissions and, as president of the Environment Council, he enjoyed the chairman's prerogative of setting the agenda.

In the event it was to take two meetings to reach the crucial agreement. At the first, in Luxembourg on 16 June, the UK agreed to sulphur dioxide emission reductions of 20, 40 and 60 per cent by 1993, 1998 and 2003 respectively, and further conceded that plants in the range of 50–100 MW be included in the directive. On the other hand, the UK pressed for a relaxation of the proposed nitrogen oxides emission targets, to 15 and 30 per cent by 1993 and 1998 respectively, instead of 25 and 40 per cent; and for an exemption from the directive if, for certain reasons, it would not be feasible to meet the emission reduction targets. It also proposed that the emission standards for new plants should be relaxed for the burning of high-sulphur indigenous coals where compliance would require very costly abatement technology. Not to be outdone, Spain sought an exemption to allow it to authorize new coal-fired power stations which did not meet the emission standards for sulphur dioxide, and France pressed for less stringent emission standards to be applied to its peak-period power stations. Still no agreement could be reached, and because several important proposals, including that on emissions from small cars, could not be properly discussed, it was decided to meet again on 28 June, just two days before the German presidency ended.

The central issue at the second meeting was a proposed directive setting stricter emission limits for passenger cars with an engine capacity below 1.4 l. This was also an issue

The Little Tennessee River

The snail darter

The Tellico Dam nearing completion

The MDPA mines near Mulhouse

A mound of waste salt from the MDPA mines

Inuit whaling captains at Barrow

The Union Carbide plant in Bhopal

Removing the centre section of the Columbiera Bridge

The first of the minehunters passing through the Columbiera Bridge

with a long history. On the table was a proposal from the European Commission for a standard known as 8 g/test for nitrogen oxides and hydrocarbons combined, which could be met by fitting vehicles with a simple oxidation catalytic converter. However, four countries – Germany, the Netherlands, Denmark and Greece – strongly supported a stricter limit of 5 g/test. To meet this standard vehicles would have to be equipped with a more expensive three-way catalytic converter and engine management system. Moreover, the UK, France, Italy and Spain – all major producers of small cars – supported a laxer standard of 12 g/test, the cheapest of the options and a standard which could be met by fitting vehicles with the so-called "lean-burn" engine. These countries were ultimately prepared to accept a compromise agreement of 8 g/test, but together they came just one vote short of a qualified majority.[3] It was therefore necessary to persuade one of the four member states supporting the limit of 5g/test to accept the Commission's proposal.

France and the UK knew that Töpfer's highest priority was the Large Combustion Plant Directive. They drew the obvious conclusion. At the instigation of France and with the support of the UK, it was made clear to Töpfer that they would agree to the latest proposal for the Large Combustion Plant Directive if Germany would drop its demand for an emission standard for small cars of 5 g/test. Töpfer's choice was to accept this deal and get both directives adopted or to reject it and leave empty-handed. He clearly had difficulties with this deal since he suggested ending the meeting six or seven times as the night wore on. It was primarily the Environment Commissioner, Stanley

---

[3]The adoption of a proposal by qualified majority in the Council of Ministers requires at least 54 of the 76 available votes. These votes are distributed amongst the member states approximately on the basis of population size, as follows: Germany, France, Italy, UK: 10 votes each; Spain: 8 votes; Belgium, Greece, the Netherlands, Portugal: 5 votes each; Denmark, Ireland: 3 votes each; Luxembourg: 2 votes

Clinton Davis, who pressed to continue the negotiations, indicating that the compromise package was also strongly supported by the European Commission. Eventually Töpfer conceded; at 4.00 in the morning of 29 June, with three countries still adamantly opposed to the limit of 8 g/test, it became necessary to vote on the proposal. The Netherlands, Denmark and Greece duly voted against, but the 63 votes of the other nine member states were sufficient to ensure the required qualified majority.

The price having been paid by Töpfer, the Large Combustion Plant Directive was then agreed with no further negotiations and without the need for a vote. Formal adoption followed on 24 November after the legal and linguistic details of the text had been finalized. After five years of acrimonious negotiations, the proposal was finally nodded through as part of a shady package deal.

The 1988 EC Large Combustion Plant Directive

The objective of the Large Combustion Plant Directive is to limit emissions of sulphur dioxide, nitrogen oxides and dust from all combustion plants with a thermal capacity of 50 MW or more, irrespective of the type of fuel used. Separate controls are laid down for existing installations and for new plants authorized after 1 July 1987.

For *existing plants*, emissions of sulphur dioxide and nitrogen oxides are to be reduced by fixed percentages in phases, as follows:

Member state	Percentage reductions in 1980 emissions				
	SO_2			NO_x	
	1993	1998	2003	1993	1998
Belgium	40	60	70	20	40
Denmark	34	56	67	3	35
France	40	60	70	20	40
Germany	40	60	70	20	40
Greece	+6	+6	+6	+94	+94
Ireland	+25	+25	+25	+79	+79
Italy	27	39	63	2	26
Luxembourg	40	50	60	20	40
Netherlands	40	60	70	20	40
Portugal	+102	+135	+79	+157	+178
Spain	0	24	37	+1	24
UK	20	40	60	15	30
EC total	23	42	58	10	30

Each member state was required to draw up a programme by 1 July 1990 setting out how these reductions are to be achieved and to submit annual emission inventories for sulphur dioxide and nitrogen oxides to the European Commission from that year. The targets may be modified by the Commission in the event of substantial and unexpected changes in energy demand, of the availability of certain fuels or of technical difficulties with generating stations.

New plants are to meet emission limit values for sulphur dioxide, nitrogen oxides and dust. For sulphur dioxide, the limit values vary on a sliding scale for plants burning solid

(continued)

or liquid fuels depending on the thermal capacity of the unit, though in each case the limit value for plants with a capacity equal to or exceeding 500 MW is fixed at 400 mg/m^3. In the case of solid fuels, limits still had to be set for plants with a capacity between 50 and 100 MW. For gaseous fuels the limits are 35 mg/m^3, though different limits apply for liquefied gas (5 mg/m^3), low calorific gases from the gasification of refinery residues, coke oven gas and blast-furnace gas (800 mg/m^3) and gas from the gasification of coal (still to be fixed).

For nitrogen oxides, the limit values are 650 mg/m^3 for solid fuels, 1300 mg/m^3 for solid fuels containing less than 10 per cent volatile compounds, 450 mg/m^3 for liquid fuels and 350 mg/m^3 for gaseous fuels.

The limit values for dust are, for solid fuels, 50 mg/m^3 for plants with a capacity less than or equal to 500 MW, 100 mg/m^3 for larger plants, 50 mg/m^3 for liquid fuels and 5 mg/m^3 for gaseous fuels, though with higher limits for blast furnace gas (10 mg/m^3) and gases produced by the steel industry (50 mg/m^3).

The European Commission is required to submit proposals to the Council of Ministers for revising the limit values for new plant by 1 July 1995 in the light of technological developments and environmental requirements.

Several derogations from the general requirements for new plants were specified. Thus, where new plants burn indigenous high-sulphur solid fuels and can only meet the emission limits by installing excessively expensive abatement equipment, fixed rates of desulphurization may instead be applied. Spain may apply rather more lenient emission controls to new plants with a capacity of 500 MW or more until 31 December 1999 in order to allow the use of high-sulphur indigenous or imported solid fuels. Rather more lenient emission standards also apply to new plants of 400 MW or more which do not operate for more than 2200 hours a year.

The directive further lays down the procedures for measuring and evaluating emissions from new plants and the methods by which the total annual emissions from new and existing plants are to be determined.

CHAPTER FIVE

ENDANGERED CULTURE VS ENDANGERED SPECIES

The Inuit and the bowhead whale

It has been claimed that an economist is someone who knows the price of everything and the value of nothing. Presumably, an economist would maintain that a commodity which has no price will therefore have no value. I cite this cynical observation not to be unkind to economists – who are only too aware of the difficulties involved in attempting to represent our intangible notion of value with the quantifiable measure of price – but to raise a fundamental environmental issue: how do we manage the exploitation of a natural resource which is "free"?

Take the case of a river. Historically, many heavy industries have been sited on a navigable waterway in order to enjoy the benefits of easy transportation and a ready supply of water. But a river also happens to be a very convenient place to dispose of industrial wastes. All a plant manager needs to do is run a pipe from the process to the river and the wastes can be disposed of at no cost to the company. Of course, those users downstream who rely on the river as a source of clean water would be forced to find an alternative supply, and the fish would not be very happy about it either. And, inevitably, it would only be a question of time before another industry upstream did exactly the same thing and proceeded to pollute our friend's own supply of clean water. Not surprisingly, it did not take very long before the general interest in maintaining clean rivers prevailed over the private interest of upstream plant owners in the form of regulatory controls on exactly how the water might be used.

River pollution is a relatively straightforward case, where the interests of the various users in a common resource are clear and their rights are readily defined – and that is why it was the first form of environmental pollution to become subject to regulatory controls. Unfortunately, many other freely exploitable natural resources are less amenable to control, and in those cases it is not unusual for exploitation to continue unrestrained, to the point where irrevocable damage is caused. Some of the most tragic of these examples concern the hunting of certain animals until they are threatened with extinction, either because the animal is regarded as a pest or because it provides a valuable product. The problem is obvious: there might well be a general interest in conserving a species in sufficient numbers to ensure a viable population – not least by the hunters themselves in order to protect their own source of supply in the long term – but it is always in the interests of the individual hunter to make one more kill. If every hunter acts on this basis, the species will ineluctably decline to the point where hunting is no longer commercially viable. And by then it may well be too late.

The most notorious case in modern times of hunting a creature to the verge of extinction is that of the whale. It has attracted attention not only because it is a classic example of the difficulties involved in managing the exploitation of a "free" natural resource, but also because the whale is a magnificent creature which needs only to surface in the vicinity of a photographer to make inroads on the public conscience. Of all the whales, it is the bowhead which is the most endangered species. An inhabitant of the arctic waters, the bowhead has been hunted by the Inuit, the proper name for Canadian Eskimos, for more than 3500 years. But it was only with the advent of commercial whaling that its numbers seriously declined, to the point where it became threatened with extinction. The reason was simple: commercial whalers operated without restriction and within a frame of reference which dictated that every whale caught represented a private gain. Exhausting the bowhead stock and destroying the basis of the Inuit whaling culture found no place on the debit side of their limited system of accounting. The problem of balancing the accounts was bequeathed to the Inuit themselves when the whalers withdrew, leaving a bowhead stock depleted to the point where it could no longer be profitably exploited. As if that were not enough, the Inuit then found that they were answerable to the international community for the way in which they managed the few bowheads which remained, with both scientific and popular opinion demanding a ban on further hunting. An excruciating dilemma arose, for the bowhead not only satisfied a major part of the Inuit's subsistence needs, it also played a central role in the cultural and religious traditions of their communities. To protect the bowhead would be to threaten the viability of the Inuit culture.

THE BOWHEAD

Whales, together with dolphins and porpoises, make up the order Cetacea. They are mammals rather than fish, having returned from the land to the sea around 70 million years ago. In size they range from the 1.5 m harbour porpoise to the 30 m blue whale, the biggest creature ever to have lived on Earth. Today, whales can de divided into two broad sub-orders: the Odontoceti (cetaceans with teeth) and the Mysticeti (cetaceans with horny plates – "baleen" – in place of teeth). One of the larger Mysticeti is the bowhead (*Balaena mysticetus*), growing up to 20 m long and weighing well over 50 tonnes when fully grown. Its name is derived from its most prominent feature, an enormous bow-shaped mouth which extends along a third of its body length. When open, the mouth reveals a vast array of baleen plates – numbering about 600 and up to 5 m long – through which the whale filters large quantities of water to trap its chief food of shrimp-like crustaceans known as krill. The absence of a dorsal fin enables the bowhead to manoeuvre directly below sheet ice, and its massive head is used to break through ice up to 60 cm thick in order to breathe. Under the black skin is a generous layer of blubber – at up to 50 cm the thickest of any whale species – which acts as insulation in the cold northern waters and provides a source of energy during winter and spring when food is scarce.

Like many other whales, the bowhead reproduces at a very slow rate. Females are thought to mature between four and ten years of age and then give birth to a single calf every two to four years. Its total numbers are unlikely ever to have exceeded 60 000, probably distributed over five geographically distinct populations around the North Pole: the Spitsbergen stock (25 000), the Davis Strait–Baffin Bay stock (6000), the Hudson Bay stock (700), the Bering Sea stock (18 000) and the Okhotsk Sea stock (6500). All these populations are closely associated with pack ice. Indeed, the bowhead is a migratory species, moving south in autumn and north in spring in tune with the advance and retreat of the polar ice sheet. It is an adaptation peculiarly suited to the arctic environment. And it was precisely this seasonal rhythm which came to be exploited by the Inuit settlers of the far north who found in the bowhead a dependable source of food and other products vital to their survival.

THE INUIT WHALING CULTURE

For much of human prehistory the Arctic remained uninhabited. Exactly when the first groups arrived is not clear, although there is some evidence indicating signs of human activity in the American Arctic in the period between 25 000

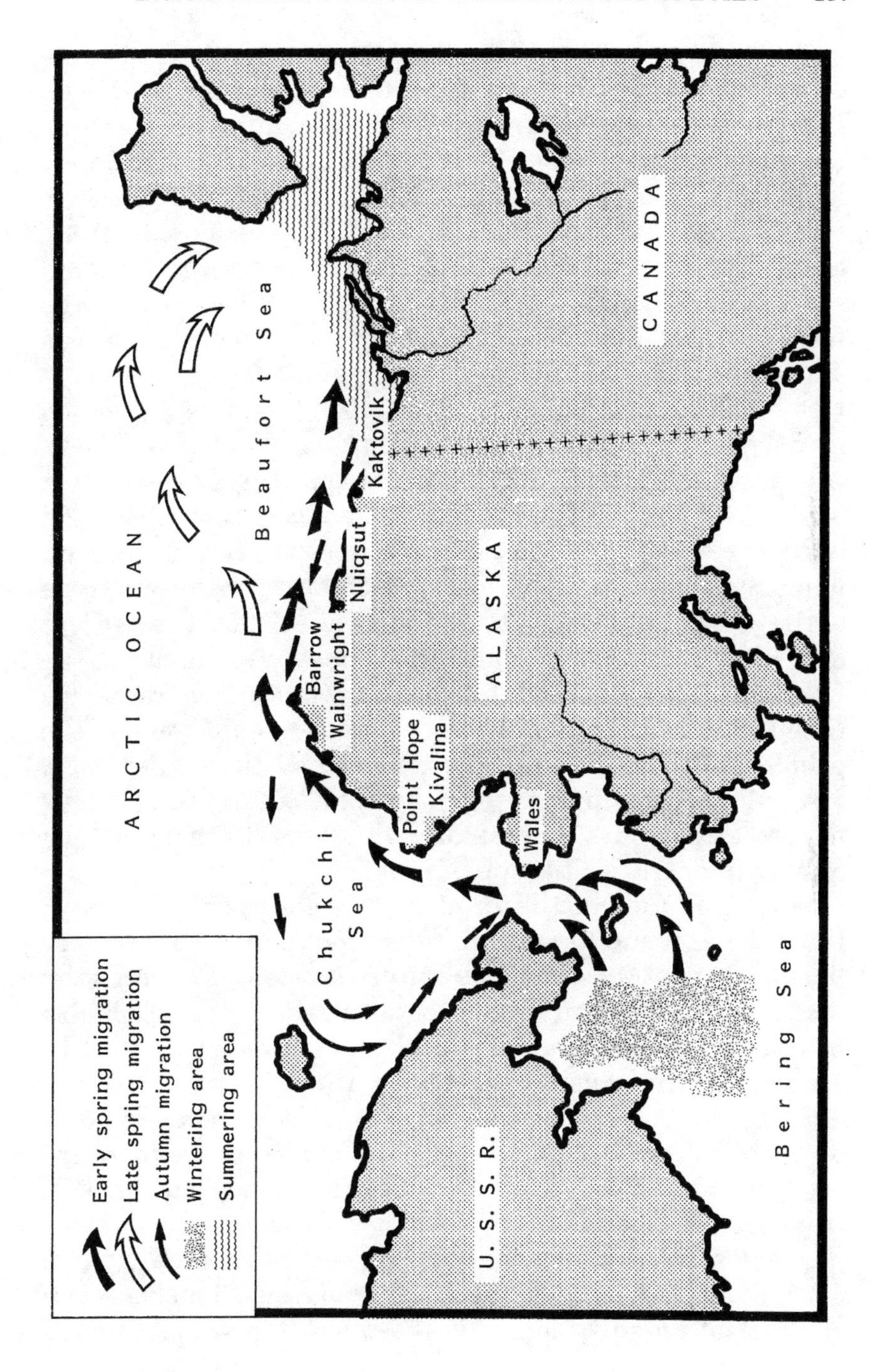
Early spring migration
Late spring migration
Autumn migration
Wintering area
Summering area
ARCTIC OCEAN
Beaufort Sea
Chukchi Sea
Bering Sea
ALASKA
CANADA
U. S. S. R.
Barrow
Wainwright
Nuiqsut
Kaktovik
Point Hope
Kivalina
Wales

and 10 000 years ago. Whether the people now known as the Inuit originated from a northern group of American Indians or migrated directly from Asia is also uncertain, although archaeological finds generally support the latter hypothesis. Not until some 5000 years ago did the first clearly identifiable culture develop. Known as the Arctic Small Tool Tradition and based on the hunting of caribou and other tundra animals, this culture spread slowly eastwards from northwestern Alaska, reaching northeastern Canada and western and northern Greenland 2000 years later. It was during this expansion, about 4000 years ago, that the first signs of a separate Inuit culture emerged in southern Alaska – the Aleutian and Southern Traditions. Other, later evidence of Inuit culture is also to be found in western and northern Alaska and the eastern American Arctic, although it must be said that the exact relationship between these early groups has yet to be established. In fact the concept of a unifying social organization has been alien to the Inuit culture right through to modern times. They recognize themselves as a single culture (the name "Inuit" means "the people") but this is primarily a device to distinguish their own culture from that of others, and particularly that of their nearest neighbours, the American Indians. To this day they remain a loosely knit group of people.

What, then, gives the Inuit culture its specific identity? Inuit are first and foremost hunters, not herders of reindeer like the Lapps. Moreover, the culture is oriented towards the sea; most Inuit groups are coastal dwellers and specialize in hunting sea mammals – seals, walruses or whales. In contrast to the popular image of the arctic way of life, most Inuit are not nomads living in igloos but are village dwellers. Their hunting lifestyle requires them to range over considerable distances in search of food, but efficient transport – dog teams in winter, skin boats in summer – enables these peregrinations to be undertaken from a permanent base. And in the absence of any extensive social or political organization, it is these isolated villages which form the focus of Inuit life; the only social units of any meaning are the immediate family, the household (which might comprise two families) and the local residence group.

Inuit have undoubtedly hunted the smaller sea mammals ever since they first settled in the Arctic, but it is unlikely that whales were hunted in the beginning, for the killing of a 50 tonne whale is a very different proposition to the harpooning of a seal. There is some evidence to suggest that whales were being taken up to 3500 years ago, but this was probably not the result of any systematic form of hunting. Whaling probably only developed into an integral part of Inuit life once population levels were high enough to permit a village to send out several crews, for the traditional type of whaling is dependent upon cooperation and manpower. Certain technological advances were also necessary, particularly the development of the toggle-headed harpoon (which, because it flexed, was far less likely to be torn out of the whale), and inflated sealskin floats attached to the harpoon line. In all likelihood it was not until around the year 800 that whaling had developed to the point where it had become a central part of the Inuit way of life in northwestern Alaska. The whale which was hunted was the bowhead, for three very good reasons: it was a slow swimmer, it was docile and it floated when dead. In fact, confronted with native hunters, the bowhead was such an exceptionally vulnerable creature that it is doubtful whether whaling could have evolved in Arctic conditions if the seas had been inhabited by a more elusive species.

This is not to say that the hunting of bowheads was a simple task; it demanded highly developed skills and was fraught with danger. Preparations began early in the year as soon as daylight returned, when all the equipment would be cleaned and repaired, the boats refitted with fresh sealskin coverings and new clothes sewn. In late March or early April, just before the first whales from the Bering Sea stock appeared during the passage to their summer feeding grounds in the Beaufort Sea, each six- or eight-man crew would retire for four days of religious preparation. The men would then depart and set up camp at the edge of the ice to await the arrival of the bowheads. When a whale was sighted, after a vigil which might have lasted several days, the crew would paddle out to the point where the animal was expected to surface again. Once struck the whale would try to escape, but the series of sealskin floats along the line slowed the animal, impeded diving and marked its position until it finally tired. Several crews might then cooperate in the kill. Any whale taken was towed ashore and, following an elaborate ritual, cut up by the entire community and distributed under the supervision of the captain. This intensive activity occupied each village throughout the spring, sometimes extending into early July. The season would then culminate in a great social and religious festival lasting several days.

To a people living in such extreme conditions any new source of sustenance is especially precious. When this source yields up to 50 tonnes of food and other essential materials with each successful kill, it is not difficult to imagine its impact on a culture. Apart from the muscle and internal organs which supplied about half the Inuit's food needs for the 40-week winter, there were the less appetizing parts which were saved for dog food; the gut was used for waterproof clothing and windows; the blubber provided oil – 20 000 l from a good-sized adult – for lighting, cooking and dyes; the baleen was made into thread, whaling gear, fishing equipment, utensils, combs, sledges, traps, spears, toys, amulets and craft objects; and the bone was utilized for harpoon heads, sledge runners, tools, fences and as a building material.

As a result of this usefulness the bowhead took on a far greater meaning to the Inuit than simply an important source of food and raw materials: it came to dominate virtually all aspects of their lifestyle, from social interaction to ritual ceremonies. Cooperation, for example, is ingrained in the Inuit culture, since survival in the harsh Arctic environment is often dependent upon sharing. Whaling reinforces this trait, demanding teamwork during the hunt and the development of social conventions for distributing the whale products to the members of the community (and even to other, non-coastal villages). Moreover, hierarchical relationships were largely a reflection of the way the whaling crews were organized, and because it was the captains of the crews who assumed responsibility for sharing out the whales taken, they came to be the most respected members of each community, being the nearest thing in Inuit culture to a leader or chief.

Given its central role in Inuit life, it is not surprising that the bowhead came to figure prominently in ritual customs, providing the theme for many festivals and ceremonies. All important social occasions, for example, are marked with the serving of *muktuk*, the blubber of the bowhead. Indeed, it was the surpluses of food generated by the spring hunt that permitted the Inuit to develop a comparatively rich artistic tradition and it would be no exaggeration to ascribe a religious significance to the creature in their culture. Certainly, the Inuit themselves consider their relationship to the bowhead to be spiritual in nature.

BOWHEAD EXPLOITATION

Despite the fact that the bowhead has been a primary element in the Inuit subsistence diet for centuries, the total numbers killed through traditional whaling have always been relatively low. Exact estimates vary, for the Inuit have no written language; legends and folklore provide the only human record. Modern research suggests that 45–60 bowheads – mostly animals less than a year old – may have been taken each year from the Bering Sea stock, although there were bound to have been wide fluctuations from year to year. It is also uncertain as to what extent hunting was carried out from sailing boats on the open sea as the bowheads migrated south in the autumn. There is, however, little disagreement that whaling on this scale posed no serious threat to the Bering Sea stock of some 18 000 bowheads.

But that stable relationship was to change suddenly and dramatically. By the seventeenth century, commercial whaling was an established industry in Europe and North America. But the bowhead, being an Arctic creature, had remained outside the range of the early land-based ships until this time. Then, in 1610, the first bowheads from the Spitsbergen stock east of Greenland were taken by northern European whalers. And it was the same characteristics which had made the bowhead the ideal target for Inuit hunting – its slowness, docility and buoyancy when dead – that were so attractive to the commercial whalers. Not only that, but each kill provided the whaler with vast amounts of oil for use as lamp fuel, as a lubricant, and for tanning and the preparation of woollen cloth; exceptional quantities of baleen were also provided, a material much in demand for corsets, hatbands, umbrellas and other products. Indeed, the whalers christened the bowhead the "Greenland right whale" for the simple reason that it was the right whale to kill. And kill it they did: within just 50 years the Spitsbergen stock of 25 000 whales had been exterminated.

Their next target was the Davis Strait–Baffin Bay stock to the west of Greenland. By the early eighteenth century whaling operations in the area were intensive, with ships from Germany, the US, Britain and especially the Netherlands actively hunting bowheads in the area; between 1721 and 1735 the number of whale-boats leaving Dutch ports each year regularly exceeded 100. In the 1750s whale-boats first began to be equipped with tryworks for rendering whale oil from blubber, thereby becoming more independent from their shore-based refining stations. It was also in this period that American colonists began exploiting the Okhotsk Sea stock in the northwestern Pacific, and in 1765 ships of the Hudson Bay's Company began hunting bowheads in Hudson Bay on a small scale, although the experiment was terminated in 1772 after just nine voyages.

But it was the nineteenth century which saw the most intense commercial whaling of the bowhead. By the 1840s the Okhotsk Sea stock was virtually extinct. Then, in 1848, the Yankee whaler Captain Thomas Roys speculatively sailed his vessel into the Bering Sea in search of new bowheads. His courage was rewarded with the discovery of the 18 000-strong Bering Sea stock. News of Roys' success spread quickly in the whaling community, and with catches declining elsewhere many fleets moved into the Bering and Chuckchi Seas to make the most of the rich pickings in the whales' summer feeding grounds. A few years later more than 200 whale-boats were operating around Bering Strait, killing about 12 000 bowheads in the following 20 years. By the last decade of the century the whalers were braving the far more dangerous waters of the eastern Beaufort Sea, overwintering on Herschel Island. Also, in 1860, American whalers resumed commercial whaling in Hudson Bay, and their early success stimulated a period of intense hunting of the local stock.

The economic motor for this flurry of activity was fuelled by the demand for oil and, above all, for baleen. With the emergence of the petroleum industry after the US Civil War, the value of whale oil progressively declined and did not in itself justify further hunting. But at the same time the price of baleen rose sharply, driven up by the fashion for narrow waists and the consequent demand for stiffened corsets. The whims of fashion provided sufficient financial incentive to persuade the whalers to endure the appalling conditions of the Western Arctic: in 60 years more than 150 ships were lost to storms, ice and shoals, and many were the whalers who, spending two or three years in the Beaufort Sea, fell victim to the bitter cold of the Arctic winter. But when a single large bowhead paid for two years at sea and one good season could make up for six poor years, many found the profits worth the risks.

The impact on the bowhead – which provided both the largest plates and the finest quality of baleen – was inevitable; by the time auxiliary steam power became widely adopted by the whaling fleets in the last quarter of the nineteenth century, thereby liberating the traditional sailing ships from the constraints of winds, currents and ice, the bowheads in the Western Arctic were in serious decline. The establishment of shore-based stations at Point Hope, Point Barrow and other locations to the north of the Bering Strait, allowing bowheads to be taken early as they swam through leads in the ice on their way to their summer feeding grounds, reduced their numbers still further. By the early years of the twentieth century, less than 3000 of the Bering Sea stock remained. This scarcity, which pushed the bowhead to the very edge of its viability as a reproducing species, forced the price of baleen even higher, making a single fully grown bowhead worth $10 000 for its baleen alone. But the very scarcity and high cost of baleen stimulated the search for a less expensive substitute material. The answer was found in spring steel, and within only a few years the market for baleen had collapsed – from $11 per kilogram in 1907 to just 17 cents in 1912. Suddenly, the economics of the bowhead whaling industry were destroyed; after 1908 the numbers of ships declined rapidly until finally, in 1915, the commercial hunt was abandoned for good.

The departure of the commercial fleets was good news not only for the bowhead; 60 years after their first rude contact with Western culture, the indigenous Inuit whalers were at last left in peace to resume their traditional way of life. The only problem was that this way of life could never be the same again. The whalers may have gone but not without bequeathing a legacy of dubious value to the Inuit. First and foremost, of course, they had taken some 20 000 Western Arctic bowheads, leaving behind a population on the verge of extinction. Faced with such a drastically depleted stock, subsistence whaling as traditionally practised could not possibly continue to support the native communities. In compensation for this, the Yankee whalers had left their technology. In the nineteenth century, many Inuit had been recruited as crew members by the commercial fleets and had been taught the Western whaling techniques. In particular they had learnt to use two crucial pieces of technology: the darting gun (which helped to incapacitate the whale when first struck) and the bomb-lance shoulder gun (which helped to kill a wounded whale). But even using these weapons, the Inuit probably took only about 10 bowheads a year after commercial whaling was abandoned, less than a quarter of their original catch.

With such a meagre catch it was clearly impossible to maintain the traditional subsistence lifestyle. Under such circumstances the Inuit were forced to become partially dependent on the American economy – as, indeed, many already had done through their contacts with the whalers. But the inevitable penetration of Western culture into the Inuit way of life proved to have insidious consequences: money came to be widely used, enabling foreign foods, materials and products to be acquired; the intensive way in which bowheads were hunted by the commercial fleets served to displace many of the rituals central to the traditional Inuit whaling culture; and with the "white man" came alcohol and disease, both of which rapidly took their toll in the vulnerable communities.

It was under these circumstances that the Inuit gradually, though only partially, recreated their traditional subsistence whaling culture. The level of hunting remained fairly constant, with about 50 crews operating from the villages between St Lawrence Island and Point Barrow and taking around 10 bowheads a year. Fur trapping provided some additional income to buy in sophisticated equipment such as outboard motors, harpoons with explosive charges and high-powered rifles. Significantly, it was the high cost of these items that effectively limited the number of crews. Had this not been the case, and had the level of predation continued at the traditional subsistence level of 45–60 bowheads a year, it is unlikely that the severely depleted Bering Sea stock could have survived. A certain stability therefore returned to the Inuit–bowhead relationship, albeit one with a distinctly precarious balance.

The problem was that it would only take a minor disturbance to throw the relationship out of balance, with possibly fatal consequences for both the Inuit culture and the bowhead. In fact the crucial event took place in the 1970s, and again it was the commercial exploitation of Arctic resources which precipitated the crisis. But before describing this new intrusion, it is necessary first to go back a quarter of a century to 2 December 1946, the date of a development which was to play a central role in the fate of the bowhead.

INTERNATIONAL REGULATION

Despite the problems caused by the unrestrained hunting of whales the world over, it was only in the 1930s that the first attempts were made to institute some system of international controls. A cautious start was made in 1931 with the concluding of the Convention for the Regulation of Whaling, a somewhat over-ambitious title for a treaty which restricted only the taking of immature and suckling whales and prohibited the commercial hunting of just a single group of whales. But that one group was the right whale (comprising the Greenland right whale – the bowhead – and the northern and southern right whales). The treaty came into force on 16 January 1935. It was, however, of limited consequence, partly due to the fact that five major whaling states – Japan, Germany, Chile, Argentina and the USSR – refused to accede to it, but mainly because the commercial hunting of the bowhead had long since been abandoned. Moreover, with no commission established to regulate the treaty, observance of the provisions by the signatory states was patchy.

With only fragmentary progress in the succeeding years, it gradually became clear, not least to the whaling industry itself, that all interests would be best served by the establishment of a comprehensive system of international controls which would provide a degree of protection sufficient to allow sustained exploitation of the various stocks. A series of conferences was duly held between 1944 and 1946 at which the framework of such a regulatory regime was worked out. The result was the International Convention for the Regulation of Whaling, signed by most of the major whaling states on 2 December 1946 in Washington, DC. Although the convention marked the first comprehensive attempt to protect whale stocks from over-exploitation, it was certainly not conceived as a species protection measure. This is made abundantly clear in the preamble, which states that the parties had agreed "to conclude a convention to provide for the proper conservation of whale stocks and thus make possible the orderly development of the whaling industry". The goal of the convention was the development of the whaling industry; the management of whale stocks at levels to support sustained exploitation was the means by which that goal was to be achieved. But if whales were not to be protected for their own sake, they were at least to be protected from over-exploitation which, in practice, was a substantial advance.

The main provisions of the convention are worth noting, not least because the way in which they were to be applied was to be of particular significance for the bowhead. The convention is binding only on the signatory states. These numbered just 15 in 1946, but have since increased to 39. All the major whaling nations are now members, although some countries which carry out coastal whaling operations, such as Portugal, remain outside the convention. As might be expected, the original signatories were all actively involved in whaling but now the majority are not whaling states; indeed, of the 18 parties which have ratified or acceded to the convention since 1979, only two maintain a fleet. Many of these states have chosen to accede to the convention with the sole objective of bringing about an end to commercial whaling, a tactic encouraged by the convention which provides that any state may become a party simply by depositing a notification in writing. This feature – not, perhaps, given much thought in 1946 – was to have far-reaching consequences 40 years later.

A special body was established to administer the convention: the International Whaling Commission, generally known as the IWC. The IWC is composed of one voting representative of each signatory state (who may be supported by experts and advisers) and is served by a full-time secretariat. It meets annually and is empowered to lay down and amend, as is appropriate, a set of regulations governing the management of the whaling industry. These regulations, known as the Schedule, may prescribe eight key aspects of whaling:

- the species to be protected;
- open and closed seasons;
- open and closed waters;
- the size limits for each species;
- the period and methods of whaling and the maximum catches;
- the types of whaling equipment;
- methods of measurement;
- the collection of statistical and biological data.

Any amendments to the Schedule are to be based on scientific findings and can only be approved with a three-quarters majority of the voting members. Interestingly, the convention prohibits the IWC from allocating catch quotas amongst the signatory states, although in practice the states themselves usually agree each year on how each quota is to be divided up (and the figures are published in the IWC's annual reports).

Much of the important preparatory work leading up to changes in the Schedule is carried out in the IWC's specialist committees. Two are of particular interest: the Scientific Committee and the Technical Committee. Both are permanent committees on which all parties are entitled to be represented. The Scientific Committee is primarily concerned with the relationship between whaling and whale stocks. Its recommendations to the full commission include, crucially, the maximum quotas which should be set for each stock on the basis of scientific evidence. The Technical Committee, by contrast, has a far broader remit, concerning itself with matters such as aboriginal whaling, killing techniques and whaling outside the jurisdiction of the convention ("pirate whaling"). Given their separate responsibilities, it is not unusual for the two committees to give conflicting advice to the IWC. Thus, while the Scientific Committee may recommend a zero quota for a severely depleted stock, the Technical Committee may declare a limited take to be justified because of its importance to a particular community.

The enforcement of the convention, as is the case with so many international treaties, is a somewhat controversial issue. The convention explicitly applies to "all waters" in which whaling is carried out by any "factory ships, land stations and whale catchers" falling within the jurisdiction of the signatory states. Two problems arise, however. First, how might contraventions of the convention's provisions be detected and what action should then be taken? And second, what is to deter a non-party which disapproves of the restrictions imposed by the convention from operating as a pirate whaler outside its scope?

Contraventions which arise during whaling operations carried out within the scope of the convention can be detected through a system of supervision which is unique in international wildlife law. Since 1949 the Schedule has required that at least two inspectors be present on each factory ship and that "adequate inspection" be maintained at each land station. These inspectors are appointed and paid by the state which has jurisdiction over the particular ship or station. In addition, since 1971 an international observer scheme has been operated by the IWC in cooperation with member nations. This was established following concern in the 1960s that some factory ships might have been taking whales in contravention of the Schedule's quotas and, after processing the carcasses, presenting them as legal catches of other species. It involves a rather curious arrangement whereby groups of whaling states agree amongst themselves to receive an observer for the purpose of overseeing their activities in a particular region – Spain and Norway in the case of the North Atlantic, for example. Some states, however, have been less than enthusiastic about this system, and in practice only four or five observers have been active.

On its own, the IWC has no power to impose sanctions on states which contravene the provisions of the convention; there is no international police force to apprehend offenders and the IWC has no authority to act as a court of law. The International Court of Justice in The Hague is recognized as the ultimate arbiter in disputes over international treaties, but the accused must first give consent to the proceedings and the court has no power to enforce its judgment. Few such cases, and certainly none concerning the International Convention for the Regulation of Whaling, have been heard by the court. There are, nevertheless, two "hard" sanctions which might be used against offending states. First, each signatory state is obliged under the convention to take action against any contraventions which occur under its jurisdiction. A list of such infractions is to be submitted each year to the IWC together with a report on the measures which have been taken against the offenders. The most common offence is the taking of undersized or lactating whales, typically punished by a fine of $1000–2000 and the forfeiting of any bonus earned by the whaler. Second, the US Congress has amended two of its fishery management statutes – the Fisherman's Protective Act 1967 and the Fishery Conservation and Management Act 1976 – so as to authorize the US government to take economic sanctions against a nation acting counter to the objectives of any international programme for the protection of endangered species in general and the International Convention for the Regulation of Whaling in particular. Given the economic importance of their trade with the US, this threat carries considerable weight in persuading many of the whaling states to comply with the requirements of the convention.

The question of how best to limit pirate whaling is, technically speaking, less easily resolved, although surprisingly it does not now seem to be a serious threat. Under its power to make recommendations to the contracting governments on any matter relating to whaling, the IWC advised signatory states in 1979 to end the import of all whale products from non-member countries and to consider prohibiting pirate whaling operations within their fishery conservation zones. These recommendations were not without effect, for many countries and the European Community have since promulgated whaling legislation. A recommendation by the IWC nevertheless remains precisely that: a recommendation.

PROTECTING THE BOWHEAD

This, then, was the regulatory framework which grew out of the realization that some form of international control was necessary if the whaling industry was to have any long-term future. As a means of protecting those stocks threatened by intensive commercial whaling, the regime had an irrefutable logic. But this logic did not always coincide with that relating to the protection of endangered species. In particular, any threat to the bowhead would come not from commercial whalers, who had abandoned it for good economic reasons, but from the Inuit who were now left with a severely depleted stock. As a non-commercial activity, aboriginal whaling had been explicitly exempted from the scope of the 1931 Convention for the Regulation of Whaling. This was done through a provision that was later to become known as the "aboriginal exemption clause": all hunting of a particular species was prohibited except where the products of the whales were intended solely for local consumption by aboriginal peoples. Fifteen years later, in 1946, the IWC opted for a similar exemption clause when drawing up the original Schedule to the International Convention for the Regulation of Whaling. Section 2 stated:

> It is forbidden to take or kill gray whales or right whales, except when the meat and products of such whales are to be used exclusively for local consumption by the aborigines.

From the IWC's point of view, there were two good reasons for adopting this strategy. First, it neatly got around the logistical problem of accumulating accurate catch data on subsistence whaling and, second, it was a logical consequence of the IWC's status as the regulatory body for and on behalf of a commercial industry; as long as subsistence whaling accounted for only a small percentage of total catches and did not threaten the interests of commercial whalers, the IWC saw no necessity to presume any jurisdiction over its control. If aboriginal whalers were to hunt an endangered (and, almost by definition, non-commercial) species to extinction – as was conceivable in the case of the bowhead – that was a matter for the aborigines concerned, not the IWC. What could not be foreseen was that a threat to an endangered species caused by continued aboriginal hunting might be widely perceived by a critical public as a failure on the part of the IWC to protect whales in general. And the possibility that such an perception might fuel a major international political issue was, in 1946, completely beyond its powers of anticipation.

To be fair to the IWC, there were few signs in the

first two decades of its existence that whaling might come to inflame passions around the world; environmentalism, after all, had yet to be invented as a political force. The new-found relationship between the Inuit and the bowhead also showed evidence of a comforting stability, the annual catch of around 10 whales following the departure of the commercial ships gradually crept up to about 15 by the early 1960s, indicating that this might be a sustainable level of exploitation. But then, 50 years after the cessation of Arctic whaling, the Western entrepreneur discovered a new and valuable natural resource in the area. This time it was oil rather than whales, and in huge quantities. But the impact on the bowhead was to be just as devastating and, worse, even more sudden.

Following the first discoveries in Prudhoe Bay, it soon became clear that the reserves of oil lying beneath the North Slope of Alaska were prodigious; known reserves were calculated at 9 billion barrels, estimated reserves at 45 billion barrels. The result was a burst of activity in the 1970s: exploration teams moved in and then, in their wake, a vast exploitation industry was built up, leading eventually to the construction of the notorious Trans-Alaskan Pipeline. Along with this economic boom came money; within just a few years a flourishing cash economy had become firmly established so that any Inuit could find employment and accumulate previously undreamt of wealth. And wealth for a male Inuit meant the ability to purchase the necessary equipment to become a whaling captain: boat, guns, bombs, harpoons, floats, lines, snowmobile, sledges, tent, stove and food, all for around $9000. Until the 1960s this scarce and valuable equipment, and with it the status of captain, was mostly passed down through families, which effectively limited the number of whaling crews. Within a decade it could simply be bought – and with inevitable results.

By the mid-1970s between two and three times as many crews were taking part in the hunt, and the number of whales taken increased accordingly, with a peak in 1976 of 48 bowheads. Not only that, but the new generation of captains lacked the hard-won experience of their forefathers. Landing a whale is an operation requiring a finely hewn sense of judgment, and all too often the whales struck by the novice captains were lost, often to die under the ice or out at sea. In 1973 10 bowheads were struck and lost, in 1977 no less than 79.

These developments did not go unnoticed in the IWC. But in considering whether to take action, it faced two difficulties. In the first place, because commercial hunting of the bowhead had been banned for 40 years, little data had been accumulated on the status of the Bering Sea population. In the absence of such information, no proper assessment could be made on whether the increased take posed a serious threat, even though this seemed likely; action required evidence of damage if it was to be approved by the whaling community. Furthermore, never before had the IWC deemed it appropriate to seek controls over aboriginal whaling – and to do so would have far-reaching ramifications for the organization, as this would effectively be seen as taking upon itself the mantle of conservation alongside its explicit task of sustainable exploitation. Nevertheless, the concern within the Scientific Committee was such that, at the 24th annual meeting of the IWC in 1972, it requested better catch data from the US on the Inuit hunt and recommended that the IWC urge the US to ensure that the numbers of whales struck and lost were reduced.

The Scientific Committee's initiative placed the US government in an awkward position. The source of this discomfiture was a recently professed opposition to commercial whaling. Traditionally, whaling had been controlled in the US through the Whaling Convention Act, which empowered the government to put the decisions of the IWC into effect. But by 1972 political support for further conservation measures was such that Congress passed the Marine Mammal Protection Act, a statute prohibiting

US citizens or US-registered vessels from engaging in whaling on the high seas. In the same year the US government also supported a proposal to the IWC calling for a ten-year moratorium on commercial whaling. Yet, at the same time as calling upon the nations of the world to end commercial whaling, it was countenancing the continued hunting, by its own citizens, of the one species of whale which was seriously threatened with extinction. And it was doing this despite having the legal powers – through either the new Marine Mammals Protection Act or the existing Endangered Species Act – to stop the hunt. Unable to resolve this dilemma, the US elected to take no action whatsoever; it neither responded to the call nor passed on the IWC's concerns to the Inuit.

The position of the Bering Sea stock became more precarious with each passing season, and the Scientific Committee reiterated its concern at each succeeding annual meeting. By 1976 research had clearly demonstrated that the Inuit whaling effort was steadily increasing, and the Committee accordingly recommended a broad programme of studies aimed at providing further information on certain crucial issues:

- a thorough examination of early whaling history, including inspection of log books to provide information on past population levels;
- marking studies to help assess mortality rates of struck but lost whales;
- an assessment of the current population status;
- the collection and compilation of better information on the sex, length, maturity and age of captured whales.

It also urged that the expansion of the fishery be limited and, again, that the loss rate of struck whales be reduced, but not at the expense of increasing the total take. For the US government, however, this call came at an unfortunate moment. In 1976, for the first time, restrictions had been imposed on an Alaskan subsistence hunt – the caribou. This animal was probably the most important subsistence species to the Inuit, and with the Western Arctic herd in decline the government had deemed it necessary to limit the numbers taken. Relations between the Inuit and the US government at that time were therefore already severely strained. An approach to the whaling captains was nevertheless made and the hunt and its impact on the bowhead population were discussed. No suggestion was made that the hunt might be either regulated or banned, although the possibility that the IWC would propose a quota of perhaps 10–12 whales a year was raised. Certainly the Inuit did not get the impression that the US supported the idea of formal limitations on the number of bowheads killed.

DOMESTIC DIFFICULTIES

The crunch came in 1977. At the 29th annual meeting in Canberra, the Scientific Committee had access to a review of published data which estimated the probable size of the Bering Sea stock in 1850 at 18 000. The best scientific evidence indicated the present size of the stock to be between 600 and 2000. With the latest catch data showing 56 bowheads killed in 1976 and 28 so far in 1977 (plus an exceptional 77 already struck and lost), the Scientific Committee registered its "real concern" as to the fate of the species. Indeed, it went considerably further, arguing that:

> any taking of bowhead whales could adversely affect the stock and contribute to preventing its eventual recovery, if in fact such recovery is still possible. . . . If a quota is set and at any time some natural disaster reduces the population to any degree, continued application of the quota will result in severe depletion and a correspondingly shorter time to extinction. . . .

> Accordingly there is a clear scientific case to be made for a moratorium on this species in the hope that it will recover to a somewhat safer level.

The Scientific Committee elected to press the scientific case: it recommended that the aboriginal exemption from the prohibition on hunting bowheads be withdrawn. In the face of such powerful reasoning, the recommendation was endorsed by the Technical Committee and then approved by the full commission, by a vote of 16–0 (though with the US abstaining). Not a single bowhead was to be taken in 1978.

Strange as it may seem, the removal of the aboriginal exemption came as a shock to the Inuit. In part this can be explained by the inactivity of the US government since 1972 in addressing the bowhead problem, despite the increasing level of concern expressed by the IWC; by no stretch of the imagination could it be said that the Inuit had been prepared for such a move. But the crux of the problem was the sudden and destructive intrusion into the Inuit way of life. To many outside observers the issue at hand was simply a question of making good the shortfall in food and resources; to the Inuit a zero quota implied cultural starvation.

The US government was now caught between two stools. On the one hand was its reputation in the international community as a leading proponent of whale conservation, a reputation which could only be safeguarded by its acceptance of a zero quota. On the other hand was its obligation to protect the rights of one of its minority peoples. Moreover, the support of the Inuit was essential if two high-profile objectives of the administration were to be met: exploitation of the huge Alaskan oil fields and the preservation of large areas of the Alaskan wilderness. President Jimmy Carter himself personified the dilemma by being both an avowed environmentalist and a champion of human rights. His position was made even more awkward by the fact that a simple and legitimate procedural device was available through which he could prevent the moratorium taking effect, for under the terms of the convention any party may avoid being bound by an amendment to the Schedule by registering an objection with the IWC within 90 days. It was a procedure which had been exploited on several occasions by whaling nations when strict catch limits had been introduced by the IWC. Ironically it had been the subject of criticism by the US which, since 1973, had urged other nations to accept the recommendations of the Technical Committee rather than to resort to the objection procedure. It was an option which Carter would probably have preferred not to have had: he was bound to be pilloried whichever decision he made.

In the event Carter elected to side with the whales and not to lodge an objection to the zero quota. Or, as the official Department of State announcement of 20 October 1977 put it:

> In order both to protect Eskimo subsistence hunting and to maintain and improve international cooperation to protect whales, the United States has decided not to present an objection at this time to a recent International Whaling Commission (IWC) action regarding bowhead whales.

The words "at this time" suggested that the administration was not as wholeheartedly committed to the zero quota as it might have been. The reaction of the Inuit community was already clear: it did not accept that the IWC should have jurisdiction over a cultural practice around which their society had revolved for a thousand years, nor that their culture should be destroyed by circumstances which had only arisen through the greed of other peoples. Indeed, the community's first response to the IWC's moratorium had already demonstrated that it was not going to take the decision lying down. In September, 70 whaling captains had come together and created the Alaskan Eskimo Whaling Commission (AEWC), a body designed to represent their interests. The aims of the AEWC were threefold:

- to ensure that the hunt be conducted in a traditional,

non-wasteful manner;
- to educate the outside world as to the subsistence and cultural importance of the bowhead;
- to promote scientific research on the bowhead so as to ensure its continued existence.

Virtually the first act of the AEWC, on the day after the Department of State's announcement, was to file a suit against the US government demanding a temporary restraining order against the government. The suit argued that, through its failure to lodge an objection to the IWC's decision, the US government had violated the federal trust responsibility toward the Inuit people and the provisions of both the Marine Mammal Protection Act and the National Environmental Policy Act. Accepting that the AEWC needed additional time in which to present a proper claim, Judge John Sirica ruled in its favour and ordered the government to file an objection to the IWC's decision. But the government immediately appealed and just three days later, 89 days into the 90-day period allowed by the IWC for objections, the Washington Circuit Court of Appeals reversed the judgment, ruling that the Inuit had not demonstrated sufficient damage to outweigh the foreign policy considerations at stake. With the deadline now imminent, the AEWC hastily applied to the Supreme Court for a stay of judgment. But, with just hours to go, Chief Justice Warren Burger denied leave for a further appeal, and so the 90 days passed without the US lodging an objection to the moratorium.

The Inuit were not to be discouraged so easily. The AEWC immediately drew up its own management scheme, impressing upon the Department of State that the hunt would continue regardless of the moratorium. This placed the government in a difficult position, for it was well aware of the practical difficulties of enforcing the zero quota. The result was a compromise proposal: instead of a complete moratorium, the US government suggested that Alaskan natives be permitted to catch a sufficient number of bowheads to meet their subsistence and cultural needs. Fortuitously, the IWC was due to hold a special meeting that December in Tokyo to discuss the controversial sperm whale quota, and the US therefore requested that the bowhead be added to the agenda. The IWC agreed, and a formal proposal that the Inuit be granted a quota for 1978 of 15 bowhead killed or 30 struck was put to the meeting. For several days no agreement could be reached. The Scientific Committee reiterated its opinion that the species might already be beyond recovery and that any further hunting could cause irreparable harm. But at midnight on the very last day of the meeting, a quota for the following year was finally approved: 12 bowheads killed or 18 struck. It was a highly controversial decision and some commentators saw a clear link with the simultaneous negotiations on the new sperm whale quota, claiming that the vote by the US representative to increase by 5681 the previous limit of 763 sperm whales for the Soviet and Japanese fleets was part of a deal to secure agreement on a bowhead quota.

But for the AEWC it was a hollow victory. Arguing that the quota was hopelessly inadequate for their needs, the Inuit representatives left the meeting in disgust. They did, however, agree to abide by the 1978 quota, pending the outcome of the government's crash research and management programme, which was carried out in close liaison with the AEWC. One of the early results of this programme was a revised (though decidedly rough–and–ready) estimate of bowhead numbers: 2264, twice the IWC's previous estimate. An assessment was also made of the cultural significance of the bowhead to the Inuit and its contribution to their subsistence needs, from which it was estimated that a minimum annual take of 24–32 whales was necessary if the basic requirements of the community were to be met.

Armed with this more optimistic evidence of bowhead numbers and the analysis of the Inuit's basic needs, the US proposed to the 1978 annual meeting of the IWC in Cambridge that the catch limit be set at the level of the aboriginal subsistence needs, provided that this did not exceed 2 per cent of the estimated stock size. At the meeting the Scientific Committee kept to its opinion that the only justifiable course of action from a biological point of view was a zero quota. The matter then passed to the Technical Committee which, after rather difficult negotiations, agreed by majority vote to recommend a quota of 24 whales landed, although it was unable to agree on the number which could be struck. The full commission found itself in similar difficulties, and it took four different amendments to the Technical Committee's recommendation before the necessary three-quarters majority could be found for a 1979 bowhead quota of 27 struck or 18 landed. Because of these problems, the IWC decided to set up a working group of the Technical Committee to examine the entire Inuit whaling issue and to draw up proposals for the Alaskan bowhead hunt in time for the next annual meeting.

This outcome still failed to satisfy the Inuit, despite indications that the US delegation had compromised its position on the sperm whale quota – essential to the Japanese and Soviet fleets – in order to secure agreement on a moderate bowhead catch limit. In fact the Inuit announced that they would ignore the IWC quota, and then proceeded to allocate themselves a total of 45 bowheads for 1979 (although in the event their hunting operations in the 1979 season were hampered by poor weather; only 22 bowheads were struck that year and just seven were landed). In July, the AEWC once more resorted to litigation, this time claiming that the regulations promulgated under the Whaling Convention Act violated the federal trust responsibility and the provisions of both the Marine Mammal Protection Act and the Endangered Species Act in that these statutes allegedly pre-empted the Whaling Convention Act with regard to the protection of the bowhead. Yet again, the Inuit were to be frustrated, for in January 1979 the court finally dismissed the case on the grounds that it lacked the jurisdiction to determine matters so closely related to US foreign relations.

INTO THE BACK ROOMS

The working group of the Technical Committee specifically set up to investigate the Inuit whaling issue concluded its work three months before the next annual meeting, held in London in July 1979. Its recommendation was for a dual system of management with, on the one hand, a major research and management programme to be undertaken by the US government and, on the other, the continued use of the convention's Schedule to limit annual catches. The working group set specific quotas of 20 bowheads taken or 27 struck in both 1980 and 1981 and higher limits of 1.0 per cent of the bowhead population taken or 1.5 per cent struck in the following years (equivalent, on the basis of the US's own survey results, to 22 and 33 whales respectively). In other words, the working group simply endorsed the strategy already adopted by the US, which was not altogether surprising as there were two US delegates in the group to every one delegate from other countries. Indeed, the US actively lobbied the other IWC member nations in the period leading up to the annual meeting with the aim of securing broad support for the working group's recommendations.

The way in which the 1979 annual meeting resolved the issue is instructive to say the least. To begin with, for the fourth successive time at an IWC meeting, the Scientific Committee recommended a zero quota on biological grounds, adding that even with a moratorium it believed the population would decline if the low estimates of gross recruitment rate were accurate. It was then the turn of the Technical Committee to consider the proposals of its working group. By now, however, it was already clear to the US that it would have to make concessions on the working group's recommendations. It was arranged that when the chairman of the committee, the Danish Commissioner Lemche, formally put forward a proposal, it would be simply for a 1980 quota of 20 bowheads taken or 27 struck. Even this went too far for the Australian Commissioner, Derek Ovington, who accused the US of resorting to short-term political expedience, characterized the proposed quota as "needless slaughter" and moved that the proposal be amended to a zero quota. This prompted the chairman to table an alternative amendment of 18 bowheads taken or 27 struck and this was put to the vote and carried by eight votes to four with nine abstentions. Those countries voting for the motion were Argentina, the US, Chile, Denmark, Japan, Mexico, South Korea and Peru. Australia, New Zealand, the Netherlands and South Africa voted against.

This was only half the battle, for if the 18/27 quota was to be approved by the full commission, it would have to be passed with a three-quarters majority. In other words, those nations which had supported the motion might expect an improved understanding in the US delegation of the needs of their own whaling fleets, while some of the nations which had opposed the motion in the Technical Committee or had abstained from voting would have to be persuaded to support the quota in the plenary session. The opportunity for these nations to strike favourable deals with the US could hardly be better, and the series of votes on other whale stocks which preceded consideration of the bowhead issue on the last evening of the meeting do indeed suggest that the exchanges which took place behind the scenes were unusually intensive.

First, despite a recommendation by the Scientific Committee that a quota of between zero and 129 be approved for male sperm whales hunted by Chile and Peru off their coastlines (no recommendation could be made for females due to poor quality data), the plenary session adopted without a vote a quota of 550 sperm whales of either sex. Then, although both the Scientific and the Technical Committee had recommended a 1980 quota of 153 for the Bryde's whale (also hunted by Chile and Peru off their coastlines), an amendment in the plenary session proposed that this be raised to 254. An abstention by the US Commissioner Dick

Frank resulted in a precise three-quarters majority. Frank also abstained on an amendment to increase the Scientific Committee's recommended quota for the fin whale stock exploited by Spain from 143 to 200. Finally, and of most significance in the longer term, a US proposal for a moratorium on all commercial whaling – reviving its stance of 1972 – failed to achieve a three-quarters majority, although to press for a world-wide ban on commercial whaling while simultaneously lobbying for a substantial bowhead quota for one's own aboriginal peoples was not, perhaps, a strategy destined for success.

By the time the bowhead issue came up for consideration it was late in the evening and all the commercial quotas had been settled. The Technical Committee's recommendation was duly presented, but the Australian Commissioner immediately proposed that the scientific evidence be respected and that the limits of 18 taken or 27 struck be reduced to zero. The response of the US commissioner was blunt: "a vote for the zero quota is a vote against the aboriginal people", and Frank stated that the US would file an objection to the zero quota if it was approved. After a heated debate the zero quota was put to the vote and defeated by eight votes to six with nine abstentions. This opened the way for a vote on the Technical Committee's recommendation, but despite additional support for the 18/27 quota from Iceland, Panama, Spain, Sweden and the Soviet Union, the Seychelles changed its abstention in the Technical Committee to a "no". The result was a final vote of 13–5, short of the necessary three-quarters majority. The Seychelles Commissioner then proposed an 18/24 quota, but no other Commissioner seconded the motion. Australia tried a 12/18 variation – the same as the 1978 quota – but the US Commissioner again threatened a formal objection, adding that the Inuit would not abide by such a low limit and that the Scientific Committee's projections of the bowhead population were, in any case, based on unreasonable assumptions and should not be taken seriously. Instead he proposed that the number of bowheads struck but lost be reduced from 27 to 26. Amazingly this proposal was approved by a vote of 12–4: the Seychelles and South Africa abstained while the Panama Commissioner, bound only by an instruction from his government to support an 18/27 quota and reasoning that an 18/26 quota was a different matter entirely, actually voted against the proposal.

The 1980 annual meeting in Brighton looked set to carry on where the 1979 negotiations left off. As far as the intensive wheeling and dealing was concerned this was undoubtedly the case. But the outcome of the bargaining, although not in itself a radical departure from previous agreements, at last provided the opportunity to develop a new approach to the dilemma. The first manoeuvre, orchestrated by the Commissioner for the Seychelles, the well-known environmentalist, Lyall Watson, was a proposal to move the aboriginal/subsistence whaling issue forward on the agenda from item 14 to item 6 so as to make it the first substantive point for discussion. The intention was to try to avoid the experience of the previous year when the bowhead take, then the last item to be voted on, was linked to, and therefore dependent on, the quotas negotiated for other whale stocks. Watson's proposal was approved by a majority of just one vote and hopes were raised that the issue might at last be decided on its merits. Yet again the Scientific Committee reiterated its recommendation for a zero quota, and this time the Technical Committee endorsed the recommendation on a majority vote, despite the presentation of a report by the US which concluded that the cultural, historic and nutritional needs of the Inuit required an annual take in the ranges 18–22, 19–33 and 32–33 bowheads respectively. However, not even a simple majority for a zero quota could be found in the plenary session, the proposal being rejected by one vote. Watson then proposed that the cultural needs of the Inuit be met by turning the hunt into a kind of sacrifice, with a single bowhead to be taken by each village. This attracted even less support, being rejected by four votes. There was now only one way out: the UK Commissioner proposed an adjournment, the motion was passed, and the wheeling and dealing could begin.

It took four days for the trade-offs to be finalized – four days of what one member of the US delegation candidly described as "sleazy deals". On the last evening of the meeting the Icelandic Chairman, Asgeirsson, formally proposed the carefully worked out compromise: a three-year aggregated quota of 45 bowheads landed or 65 struck, with a maximum take in any one year of 17 whales. This was duly passed by a vote of 16–3 with five abstentions. By then, of course, all the other sensitive issues had been settled, including several quotas which comfortably exceeded the recommendations of the Scientific Committee, in particular those for sperm whales, minke whales and fin whales. The proposal for a complete moratorium on commercial whaling again failed to win a three-quarters majority, as did a proposal for a three-year moratorium for sperm whales only, although only by a single vote. It was on these issues that the new bowhead compromise was to have a telling effect. At first glance the three-year quota gave the impression of simply consolidating the aboriginal hunt, but in practice it served to remove the issue from the IWC agenda for the coming two years, thereby leaving the US free to take a tougher line on the other major issues, which it could not do when it knew that the only way of securing a substantial quota for the Inuit was to compromise on other proposals.

Seen in this light, the 1981 and 1982 annual meetings were crucial to whale conservationists, for not only was it a golden opportunity to agree strict controls on all commercial whaling, but progress here would free the agenda in 1983 for a renewed attempt at resolving the bowhead issue. The 1981 meeting, also held in Brighton, saw no less than five substantive broad-ranging proposals for controlling whaling:

- a UK proposal for an indefinite moratorium on all commercial whaling;
- a UK proposal for an indefinite moratorium on commercial whaling in the North Atlantic;
- a French proposal for a ban on the taking of minke whales by factory ships;
- a joint French/Dutch/Seychelles/British proposal for a zero quota for sperm whales;
- an Australian proposal for a phased move towards a world-wide ban on commercial whaling.

The pressure for strict controls was clear to all; indeed, each of the five proposals was passed by a majority of four or more votes. But only the zero quota for sperm whales received the necessary three-quarters majority, and even this was secured by a concession to Japan whereby the coastal catch limits in the western North Pacific area were to remain undetermined for an unspecified period. With only marginal progress on these proposals, it was perhaps fortunate that a bowhead quota was not on the agenda, for the Scientific Committee registered its concern at the fact that the 1980 catch limit had been exceeded by eight strikes and reaffirmed its view that to reduce the risk of extinction no whales should be taken. Negotiating a new quota under such circumstances would not have been an easy task.

CLOSED FOR BUSINESS

If the bowhead hunt remained a contentious issue, a general moratorium on commercial whaling was clearly an idea whose time had come. Thus it was that in 1982 the IWC took the most radical decision in its 36-year history. On the first day of the 34th annual meeting, once more held in Brighton, the proposal to set the catch limits of all commercial stocks at zero, commencing in 1984, was once again on the agenda. This time the Technical Committee supported the proposal by a larger majority, of 13. However, the nine abstaining members included representatives from several prominent whaling nations, such as Brazil and Iceland, who might decline to support such a hard-line motion in the plenary session. The proponents of a moratorium, led by the Seychelles delegate Lyall Watson, were therefore careful to incorporate two concessions into the final proposal: the moratorium was to commence after three years instead of two and, by 1990 at the latest, the IWC was to undertake a "comprehensive reassessment of the effects of this decision on whale stocks and consider modification of this provision and the establishment of other catch limits". On the basis of this reassessment the moratorium could be reviewed. The tactic worked, and the proposal was adopted by a vote of 25–7. At last the conservationists had achieved a ban on commercial whaling; although formally only a moratorium, it was a decision which would prove difficult to reverse, for the three-quarters majority necessary to amend the Schedule – for so long an unnegotiable obstacle to conservationists – now served as a serious impediment to the resumption of commercial whaling.[1]

The moratorium was welcomed as a triumph by the conservationists. But, ironically, it only complicated the bowhead issue and caused acute embarrassment to the Inuit, for while all commercial stocks were to be – at

---

[1]Although not directly relevant to the bowhead issue, it deserves mention that the moratorium proved to be far less effective than had been hoped. Japan, Norway, Peru and the Soviet Union formally objected to the zero catch limits with the result that they were not bound by the decision. Furthermore, some traditional whaling nations, such as Iceland and South Korea, adopted, on a large scale, the dubious practice of "scientific whaling". Under the convention any country may award itself a permit to take whales "for purposes of scientific research". The whale meat so obtained may be used for human consumption. Whether the taking of whales is necessary for the advancement of scientific understanding of the various stocks is a contentious issue. Most independent scientists maintain that virtually all the data required for the management of stocks can be obtained without killing any whales.

least officially – fully protected, the most endangered of all whale species was still being hunted. The only consolation for the Inuit was that the size of the Bering Sea stock was now estimated by a special sub-committee of the Scientific Committee at about 3857 individuals, around 1500 more than the estimates made in previous years and between 21 and 43 per cent of the initial population. But the Scientific Committee then proceeded to throw a damper on the situation by adding that it could not decide whether there had been any net recruitment to the population since 1915. It therefore maintained its long-standing view that the safest course for the recovery of the stock was a zero take but, noting that a three-year quota was in force, recommended that only sexually immature specimens be taken so as to leave the stock of reproducing adults intact.

The three-year quota was due to end in 1983, and the parties were soon taking up their positions for the approaching battle. Even before the exchanges got under way in Brighton, the new US commissioner, John Byrne, had entered into an agreement with the AEWC that he would aim for a quota of 35 strikes – a move which, predictably, angered the conservationists. The Scientific Committee also generated some controversy: rather than pressing once more for a zero quota, it recommended instead that "extreme caution" be taken in setting catch limits and requested that, if bowheads were to be taken, the catch limit for 1984 be less than 22 and that these be sexually immature animals. The Technical Committee failed to be persuaded by the US plea for a one-year quota of 35 strikes or 26 landed (which was hardly surprising since Byrne was to admit that the figures were based on "quick and dirty" estimates), adopting instead a proposal from the Netherlands for another three-year quota, this time for a total of 42 strikes with a maximum in any one year of 10 whales landed. Neither this nor any of four other proposals managed to secure a three-quarters majority in the plenary session, and only after lengthy back-room discussions could a suitable compromise be patched together: a two-year quota of 45 whales struck with a maximum number of strikes in any one year set at 27, with the proviso that the total number of strikes be reviewed after the first year. For the first time no distinction was made between strikes and kills – a considerable incentive to the Inuit to improve the effectiveness of their hunting methods.

With a two-year quota again in force and no significant scientific developments, 1984 proved to be a quiet year for bowhead followers. But it was business as usual in 1985. When the delegates met at Bournemouth, the first point under agenda item 13 on aboriginal/subsistence whaling was a lengthy report from the Scientific Committee which presented a reassessment of the bowhead stock size. On the basis of visual censuses it was concluded with a 95 per cent probability that the population numbered between 3165 and 3711 in 1978 (with a mean of 3440) and between 3222 and 4538 in 1982 (with a mean of 3880). However, when account was also taken of recent aerial surveys and acoustic censuses, the range for 1984 came out at a sloppy 2613–6221 (with a mean of 4417). Although the figures seemed to be indicating a steady recovery in the bowhead population, the fact is that bowheads simply cannot breed as quickly as the trend was suggesting; indeed, virtually any interpretation could be placed on the statistics to suit the purpose at hand. The only conclusion which could be drawn was that the surveys gave a greater insight into the limitations of the measurement techniques themselves than into any likely changes in the bowhead population.

The Scientific Committee also reported that the number of bowheads struck in 1983 and 1984 totalled 43, two less than the quota had allowed, and that the struck–and–lost rate had improved from 52 per cent in 1983 to 38 per cent in 1984. It nevertheless recommended continued caution in setting any new catch limits. In the Technical Committee, the US emphasized earlier work which indicated that the need of the Inuit community was 35 strikes a year, but a Finnish proposal of 43 strikes over two years, with a maximum in any one year of 27, was adopted by a majority, with agreement on the possibility of review and amendment after one year. In the plenary session, however, no two-thirds majority could be found for either this or an Irish amendment of 50 strikes over two years; US commissioner John Byrne reiterated his plea for 35 strikes a year and Mexico reminded the gathering of the Scientific Committee's cautionary view. The inevitable adjournment followed, and further hard bargaining eventually produced a new consensus: a three-year quota of 26 strikes a year with strikes not used in any one year being transferred to the next year to allow a maximum of 32 strikes. A footnote providing for a review of the limits each year was added, although any amendments would have to be based on the advice of the Scientific Committee.

This agreement provided the opportunity for a degree of peace to settle once more over the bowhead issue and gave the scientists a further three years to carry out more extensive research on the bowhead population. Thus, by the time the bowhead quota again came up for consideration at the 1988 meeting, further surveys of the bowhead population had been made. These surveys had the effect of relieving some of the pressure on the delegates, for the analyses produced a new, higher estimate of of 7800 bowheads, with a confidence interval of 5700 to 10 600. (Subsequent research, using new information on the swimming behaviour of the whales and acoustic location techniques, has broadly supported this estimate.) The consequence of the more optimistic population count was a willingness by many delegates to accept a higher three-year quota, and agreement was reached on a limit of 44 strikes or 41 whales landed for each year in the period 1989–91, with a provision to allow up to three unused strikes to be transferred to the following year.

THE UNENVIABLE PROSPECT

For a resource manager the bowhead issue has all the characteristics of a zero-sum game: any gain to the whale in the sense of a lower quota represents a corresponding loss to the Inuit culture; any concession to the whalers through higher catch limits increases the threat to the bowhead. Finding a sustainable balance between the two interests is an unenviable task. In fact, such a balance may not even exist; present bowhead numbers may be so low that *any* level of hunting would prevent recovery to a visible breeding population. Both the Inuit culture and the bowhead species might already be doomed to extinction: the Inuit culture because the Western influence has already largely displaced the traditional hunting lifestyle; the whale because its numbers might have fallen below the minimum number necessary for the population to sustain itself – or at least to be able to withstand a single catastophic event, such as a pollution incident – and because the average age of the population may have become dangerously high.

That the Inuit cannot be blamed for the dilemma in which they find themselves is of little consolation. As much as the children of the society which bore the true responsibility for creating the dilemma might wish to ease their consciences by taking an active role in the resolution of the issue, the Inuit culture has already suffered enough from external influences. Then again, it might only be these children of conscience who are in a position to hold back the ultimate threat to both the bowhead and the Inuit. Of all the US's indigenous reserves of oil and gas, some 40 per cent are to be found in the Beaufort Sea, directly beneath the summer feeding grounds of the bowhead. The North Slope of Alaska is steadily being opened up to exploration for oil, gas and minerals and the extraction industries are ready and waiting in the wings. Within such a scenario, the presence of an endangered species is a serious complication. For a creature already on the verge of extinction the prospects would be dire indeed; intense human and industrial activities would disrupt its habitat and a major oil spill could effectively spell its doom. The prospects for the Inuit culture, if the history of aboriginal peoples confronted by Western capital is anything to go by, are equally ominous. For, while it remains uncertain whether both the bowhead and the Inuit can prosper together, there is no doubt at all that both species and culture could perish together.

CHAPTER SIX

A CERTAIN ACCIDENT

The tragedy of Bhopal

Suppose you are a government official, responsible for industrial safety, in a Third World country. You are confronted with a structural problem. The regional economy is growing steadily but development is taking place in an uncontrolled fashion: industrial growth is concentrated in relatively few cities; there is a virtual absence of rural development; and the physical and social infrastructure is only in its formative stages. You are daily confronted with the problem of a society that encourages the rapid development of industry but which has only limited technical, administrative and financial resources for regulating industrial activities.

A large chemical plant in one of your region's most important industrial centres has for several years produced a range of pesticides. The component chemicals needed to do this have been imported from other suppliers. Competitive pressures, however, have persuaded the company to expand its plant enabling it to produce the component chemicals itself. One of the chemicals involved is highly toxic and unstable, and its manufacture involves an extremely hazardous process. The operation of the new process contravenes the local development plan and the site is located in a densely populated part of the city. But the plant provides work for hundreds of people and pays relatively high wages. It is also operated by one of the largest – and most influential – companies in the country.

You would like to carry out a proper assessment of the hazards associated with the plant but your control agencies have neither the resources nor the expertise to carry out a comprehensive risk analysis. Does the international standing of the company provide you with sufficient confidence to countenance the new process, or do you insist on strictly enforcing the law and threatening the plant with closure, thereby making a large number of workers unemployed and damaging the much-needed industrial development of the area?

THE SETTING

India is second only to China as the most populous country in the world, but with a gross national product per capita of less than $300 it is also one of the poorest. Compared with many other Third World countries, however, India is home to a wide range of industries and has invested substantial resources in developing its own scientific and technological capability – although, like so many other developing countries, the benefits of industrialization tend to be distributed unevenly, both geographically and socially.

Bhopal, the state capital of Madhya Pradesh, is a good example of the extremes which are to be found in developing countries. Although a thousand years old, Bhopal acquired an industrial base only comparatively recently when, in 1956, it was designated as the state capital. The city quickly developed a strong public sector and within a relatively short time the government had become the area's biggest employer. Large-scale industrialization began in 1959 when Bharat Heavy Electrical Ltd located a huge plant for the manufacture of electrical equipment just outside the city boundary. More than 50 000 people were employed on the site and supply industries and other enterprises quickly followed. Almost overnight Bhopal had become a major industrial centre.

In terms of development, Bhopal is certainly more fortunate than many other towns and cities in India. But the physical infrastructure suffers from serious shortcomings, so that housing, transportation and communication facilities are poorly developed and supplies of water and energy are limited and unreliable. Public services, such as health care and regulatory activities, are similarly rudimentary. Many of these shortcomings are simply due to the dynamics of Third World development. The immediate response to industrial growth in a city such as Bhopal is a mass migration of people from the surrounding rural areas; in the 1970s the population of Bhopal grew at a rate three times that of the state as a whole. Adequate housing was simply not available, and land and construction costs rose as demand rapidly outstripped supply. The inevitable result – seen in similar situations throughout the Third World – was the uncontrolled growth of shanty-towns. By 1984, 130 000 of the city's 700 000 inhabitants lived in these slum colonies, mostly located close to the sites which provided employment for the cheap labour.

To plan for ordered development in such circumstances is an unenviable task. On two occasions the city authorities drew up broad development plans – the Capital Projects Development Plan of 1958–59 and the Interim Development Plan of 1962–63 – but the rapid influx of people and the anarchic expansion of industrial activities defeated the planners' efforts. The results of this failure were to prove fatal.

THE PLANT

In 1934 the US Union Carbide Corporation registered a private company in Calcutta for the purpose of assembling batteries. Union Carbide had started operations in 1886, producing carbon products, but later diversified into industrial gases, chemicals, petrochemicals, consumer products, speciality products and technical services, including the operation of nuclear power plants. It gradually developed into one of the biggest chemical companies in the world, employing 100 000 people in 40 countries and with annual sales of around $10 billion. Its Indian subsidiary, Union Carbide (India) Ltd, also grew steadily, expanding its operations to become the twenty-first largest company in India with more than 10 000 employees and annual revenues of about $170 million. Its US parent retained a majority shareholding of 51 per cent.

One of the sectors into which the Indian subsidiary diversified was pesticides. Pest control was seen as a prerequisite for the success of the "Green Revolution" and, with the market for pesticides growing rapidly in the 1960s, Union Carbide took the logical step of establishing an Agricultural Products Division. The venture started modestly with the opening of an office in Bombay in 1966. In 1968 the division moved to a new formulation plant in Bhopal where, starting in 1969, a number of different preparations were produced using carbaryl as the main active agent (also known by its trade name "Sevin"). At that time the plant had no capacity to manufacture the constituent substances; the necessary agents were imported from other manufacturers and mixed to create the appropriate pesticides.

In light of later events, it is worth emphasizing that the new plant was located outside the city, about 2 km from the central railway station. It was only due to subsequent growth, both of the installations and of Bhopal, that the site became absorbed into the city. The site itself originally covered an area of about 2 ha, but in subsequent years the expansion of the original facilities and the addition of a wide range of new processes led to the steady growth of the plant until it extended over 30 ha. It was in this same period, stimulated by the growing industrial and commercial activity in Bhopal, that the population of the city more than doubled, from 300 000 to 700 000 inhabitants. The only place for many of the new inhabitants to live was in the slum colonies which quickly appropriated any

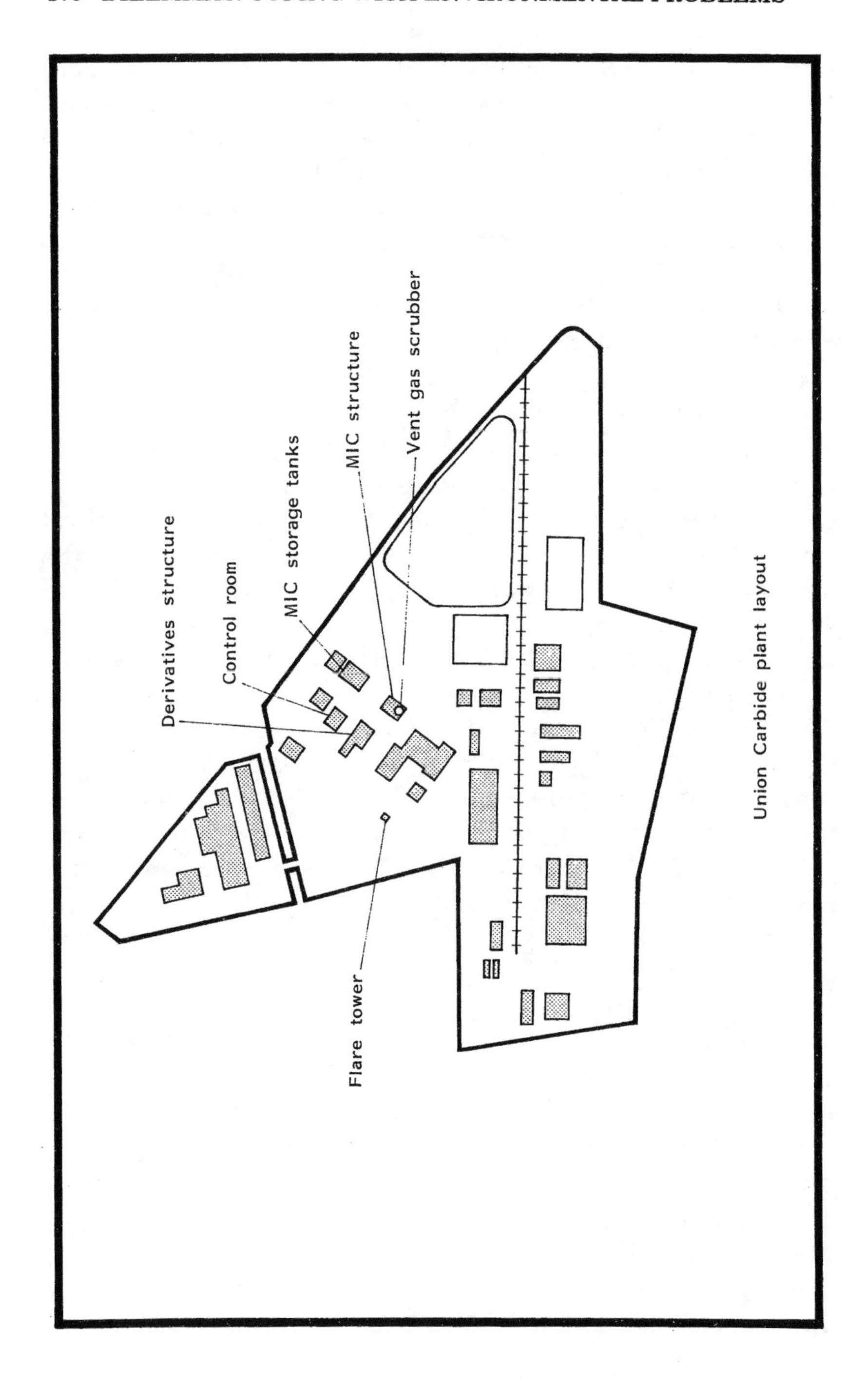

Union Carbide plant layout

available open land. Two of these colonies, Jaya Prakash Nagar and Chola Kenchi, grew up on the perimeter of the Union Carbide plant.

It became clear early in the life of the Agricultural Products Division that buying in the agents for its products from other suppliers was a relatively expensive way of producing pesticides. Union Carbide therefore drew up plans to develop a manufacturing capability at the Bhopal site. Accordingly, in 1970, shortly after the plant became operational, the company applied to the central government for a licence to manufacture pesticides. Discussions were also held with the US parent company on obtaining technical assistance for the new venture. These discussions resulted, in November 1973, in a $20 million agreement between the Union Carbide Corporation and Union Carbide (India) Ltd for the transfer of equipment and technical know-how for the manufacture of pesticides based on methyl isocyanate.

THE CHEMISTRY

Methyl isocyanate (MIC) is one of several isocyanates, a group of molecules which are characterized by their propensity to react with compounds containing active hydrogen atoms. The isocyanates are an example of "cumulated unsaturated systems" – molecules which end in a double adjacent bond of carbon and oxygen or carbon and sulphur and are therefore highly unstable, reacting vigorously with other organic compounds. It is this reactivity which makes the isocyanates both useful and hazardous. Three compounds from the group are used widely in the chemical industry: toluene diisocyanate (TDI), 4,4'-diphenylmethane diisocyanate (MDI) – both used in the polyurethane industry to make foams, elastomers, insulation and coatings – and MIC, which is used almost exclusively for the production of carbomate pesticides.

Of the three chemicals, MIC is the most hazardous. In

the first place, MIC is highly volatile, boiling at 39°C, and the vapour is approximately twice as heavy as air. It is highly reactive, having an affinity to a wide range of compounds, and it is also flammable. But most importantly it is extremely toxic: the maximum allowable concentration prescribed by the US Occupational Safety and Health Administration for work exposures is the extremely low value of 0.02 ppm. Exposure to MIC – by inhalation, swallowing or skin contact – leads quickly to a reddening of the eyes and skin, poor vision, pain in the eyes and lungs, coughing and difficulty in breathing, followed by vomiting, diarrhoea and stomach cramp. In serious cases of inhalation, damage caused to the walls of the lungs leads to the release of fluid which will accumulate and eventually cause death by suffocation. Spasmodic constrictions of the bronchial tubes might also cause suffocation. However, these extreme symptoms do not become apparent until some hours after exposure, which can deceive those concerned into believing that the poisoning has not been acute.

These qualities mean that MIC has to be handled with the utmost caution. Containers and equipment must be glass-lined or manufactured from stainless steel. They should also be oversized to allow for expansion. Hoses must be of stainless steel or lined with fluorocarbon resins. Whenever contact with air is possible, such as during transfer operations, MIC should be cooled with dry nitrogen. All personnel in the vicinity should then wear full protective clothing equipped with breathing apparatus. When stored in bulk, MIC should be cooled to about 0°C as this slows down, but does not eliminate, any reactions and therefore allows more time in which to initiate remedial action. Any reactions which occur at ambient temperatures tend to be vigorous and to generate large quantities of heat, sometimes to the extent of being explosive.

One of the most likely contaminants is water. A reaction between MIC and water produces methylamine and carbon dioxide; the methylamine then reacts further with MIC and other reaction products to produce either 1,3-dimethylurea (in the presence of excess water) or 1,3,5-trimethylbiuret (where excess MIC is present). Unless the heat is removed, the temperature of the mixture will quickly increase to the point where the MIC starts to boil violently. If the reaction takes place in a closed container, the rapid build up of pressure will cause the relief valves to open and lead to the venting of the gases or, in extreme cases, the rupture of the vessel.

THE MIC PRODUCTION FACILITY

The first drum of MIC, intended for experimental purposes, was shipped by Union Carbide from its US plant to Bhopal in November 1973. Shortly after, the Indian government granted the company a licence to manufacture (rather than simply formulate) pesticides based on carbaryl, but it was not until 1977 that the new equipment was installed and production commenced. This, however, was not to be the final configuration of the plant. The increasing competitiveness of the pesticides market persuaded Union Carbide to exploit economies of scale and to reduce transportation costs by constructing production facilities for five of the chemical components which it was still purchasing from other suppliers.

Carbaryl can be produced in several ways. When it first started production of the pesticide in 1958, Union Carbide used a process in which α-naphthol was reacted with phosgene to form 1-naphthyl chloroformate and hydrogen chloride; the chloroformate was then reacted with methylamine to form carbaryl, leaving some residual hydrogen chloride. In 1973, however, the company changed to a different process in which MIC was simply mixed with the α-naphthol to form the carbaryl. The advantage of this method was that no by-products of any consequence were generated, enabling Union Carbide to supply MIC to

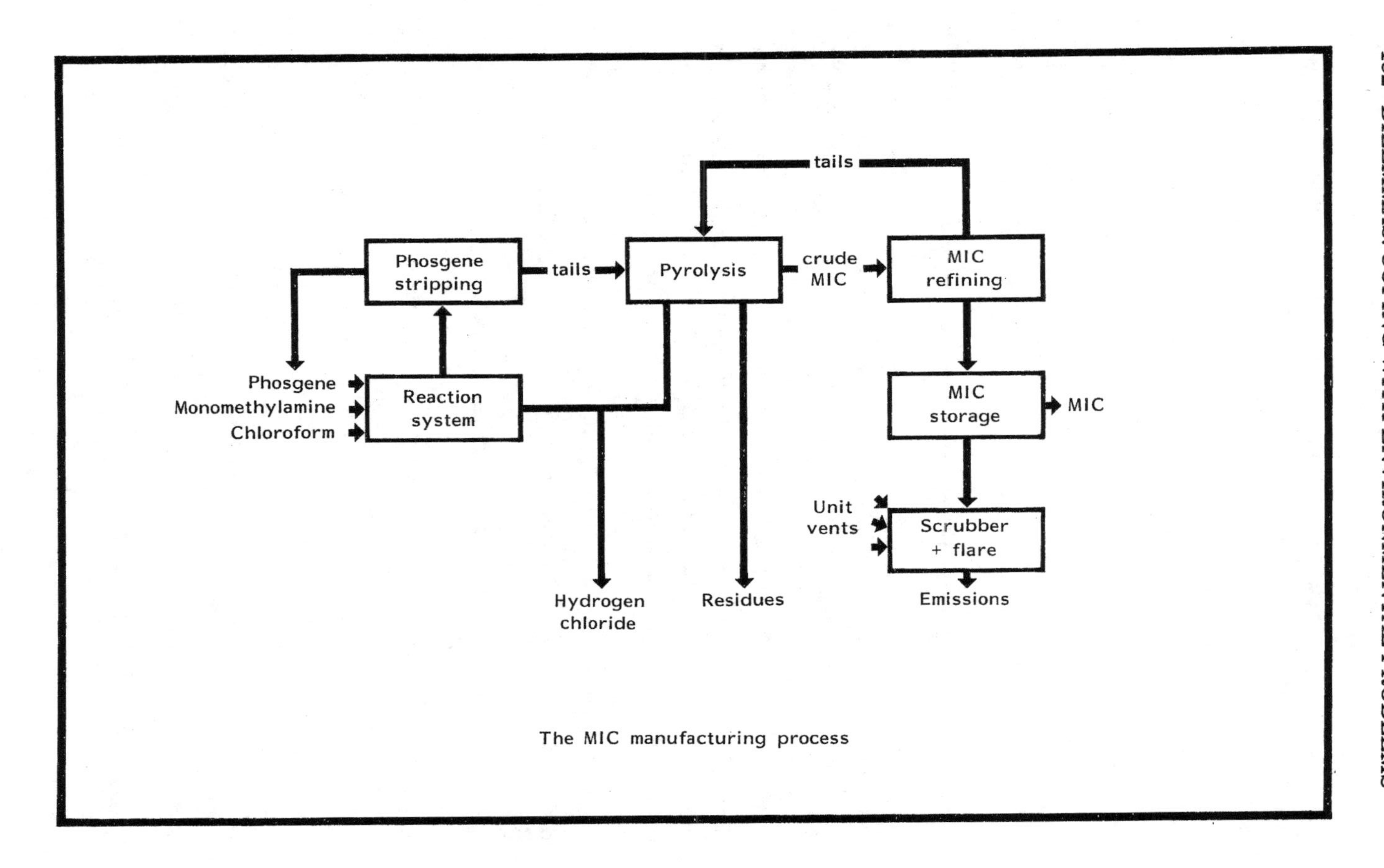

The MIC manufacturing process

other pesticide manufacturers such as Du Pont and FMC. But it also required Union Carbide to manufacture the MIC, although in itself this is relatively straightforward as chemical processes go. Monomethylamine (MMA) and phosgene are mixed and then heated to form MIC and hydrogen chloride; the products are then chilled and the gaseous hydrogen chloride is separated from the MIC in an absorber.

The basic requirements for the manufacture of MIC are, therefore, supplies of phosgene and MMA. Phosgene was produced on site by reacting chlorine (also manufactured at the plant) and carbon monoxide, but both the chlorine and the MMA were brought in from other suppliers in India. The phosgene and the MMA were then converted into methylcarbamoyl chloride (MCC) and hydrogen chloride in a vapour-phase reaction system. Additional phosgene was added to ensure that all the MMA was converted before quenching the reaction with chloroform. The mixture was fed first to the phosgene stripping sill, where the unreacted phosgene was removed and recycled, and then subjected to pyrolysis to separate the MIC from the hydrogen chloride.

At this stage the MIC is still in a crude form and has to be separated from the chloroform, MCC and other residues in a refining still. Thereafter the MIC is put into bulk storage to await its transfer to the pesticide production units. Storage at the plant was provided by three 68 000 l tanks, manufactured from 304 stainless steel and cooled to a temperature of 0°C. Only two of the tanks were used at any one time, the third being reserved for emergency use or for the temporary storage of off-specification MIC pending its reprocessing. Each tank was purged under pressure with dry nitrogen and fitted with a high-temperature indicator and alarm, a pressure indicator and controller, a level indicator and alarm, and a rupture disc designed to blow when pressure in the tank reached a critical level. Any gas escaping from the tanks, either routine discharges of nitrogen or an emergency high-pressure discharge of MIC,

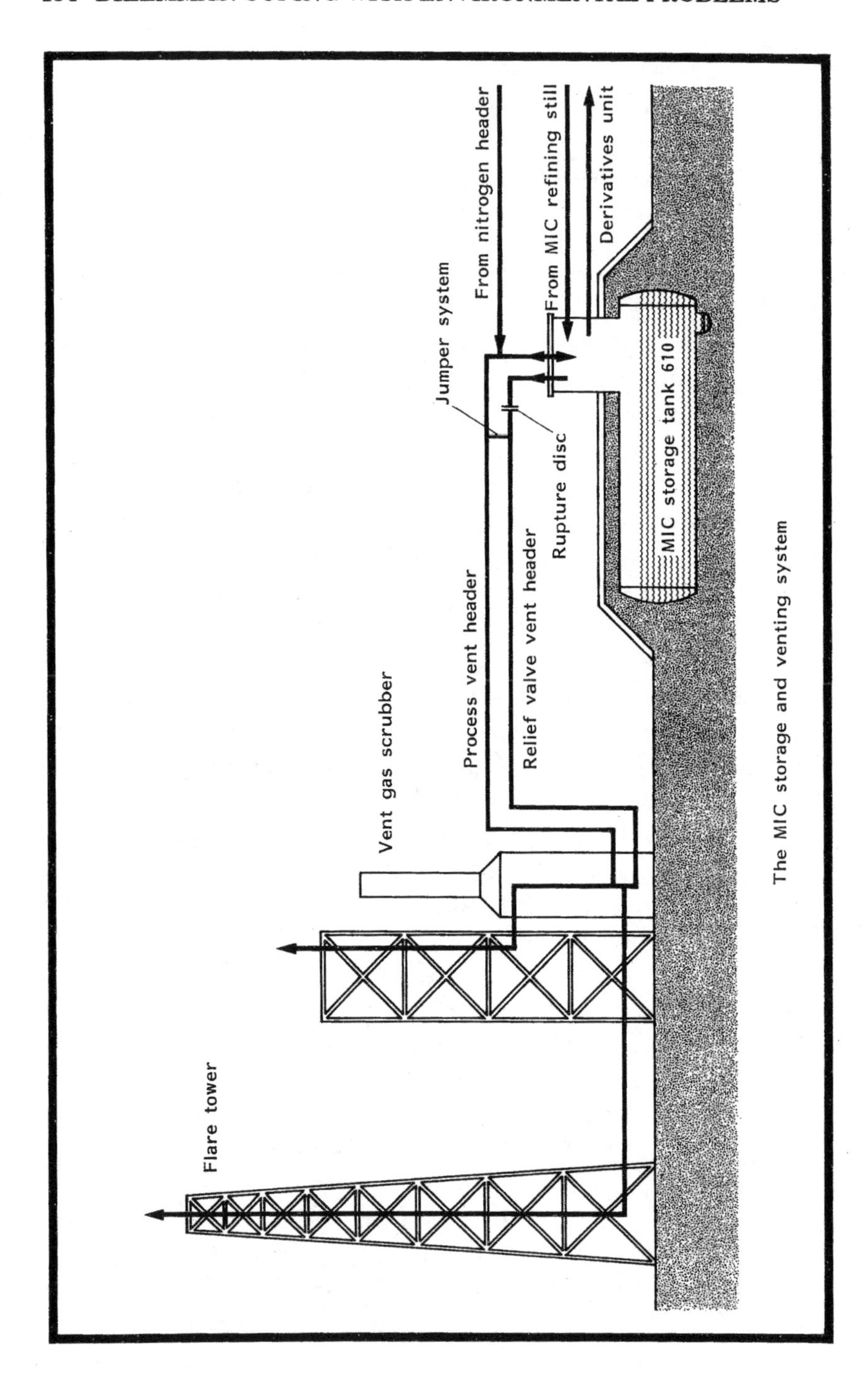

The MIC storage and venting system

was fed into a vent-gas scrubber – a 30 m high tower which neutralized escaping gases with a solution of caustic soda. The escaping gases could also be routed to a flare, either before or after passage through the scrubber, but this was not designed to destroy large quantities of MIC vapours.

The technical expertise necessary to construct the MIC-production facility was not a serious problem for the Indian subsidiary since it could call on the experience of the US parent company's plant at Institute, West Virginia. According to Union Carbide, the specifications of the facility were laid down by the parent company; detailed design and construction were the responsibility of Union Carbide (India) Ltd, using local materials and equipment where possible.

There were, however, important differences in the design and construction of the two facilities. Most importantly, the Bhopal plant incorporated a greater number of manual controls, including those operating the safety devices. Union Carbide claims that the Indian government preferred a more labour-intensive plant for employment reasons, but denies that this resulted in any lowering of safety standards. The fact remains, however, that the MIC manufacturing process involved significantly greater hazards than the operations previously carried out on the site. Indeed, it was because of the increasingly hazardous processes on the site that the municipal authorities objected to the continued operation of the plant. Bhopal's 1975 development plan had zoned the site for commercial or light industrial use, but the steadily increasing manufacturing capability of the plant clearly gave it the character of a heavy industrial activity. However, closing down recently installed capacity or moving the plant to a new location was hardly an economically viable proposition at such a late phase in the company's expansion programme. Union Carbide (India) Ltd was also a major, high-paying employer, and its influence ensured that both the state and central governments overruled the city's objections and authorized the operation of the new processes.

THE PRELUDE

The new MIC facility became operational in December 1979. Thereafter production increased rapidly, and in 1981 the output of MIC-based pesticides had reached 2700 tonnes. Unfortunately for Union Carbide, however, market conditions for the sale of pesticides deteriorated sharply during this period. There were several reasons for this: agricultural production in India had peaked in 1979 and then dropped sharply in 1980; poor weather in 1982 and 1983 led to a reduction in the use of pesticides; and new imported products such as synthetic pyrethroids increased the competition in a market already suffering from large unsold stocks and under-utilized production capacity. Union Carbide's production of MIC-based pesticides therefore declined in the years after 1981 – to 2300 tonnes in 1982 and 1650 tonnes in 1983.

To some extent, these market difficulties were a confirmation of the doubts which had been expressed within Union Carbide when the pesticides manufacturing plant was under construction. Even then there were clear signs that the pesticides market was declining and suffering from overcapacity, but it was decided that the project was too far advanced to abandon. Given the difficult commercial position of the plant and the fact that both it and the Agricultural Products Division were relatively unimportant to Union Carbide in strategic terms, the local management found it difficult to compete for resources against other divisions of the company. One important consequence was that the plant was not well positioned to attract high-quality leadership; between 1969 and 1984, no less than eight different managers were employed.

Union Carbide carried out a detailed safety audit of the entire plant in 1982. The investigation brought to light a number of problems, both serious and less serious, although it was concluded that none posed an imminent danger or required immediate correction. In the light of later events, a number of the shortcomings set out in the report are worth emphasizing. It was noted, for example, that there was a risk of an accidental release from both the phosgene and the MIC units caused by equipment failure, operating errors or inadequate maintenance. The possibility of the contamination, overpressurizing or overfilling of the MIC storage tanks was also noted, as were the absence of water-spraying equipment in certain parts of the plant (required to neutralize escaping toxic gases) and problems concerning the reliability of safety valves and the instrument-maintenance programmes.

A general observation of the safety audit was that operations were hindered by the high turnover of personnel. This was largely due to the low morale of employees, itself a consequence of the fact that the plant was running at a loss – with all that this implied for future career prospects and management–worker relations. The result was slack discipline amongst the plant personnel. This had particular impact upon the way in which safety precautions were enforced; for example, maintenance staff worked in prohibited areas without formal authorization, and the fire-watch attendant was obliged to perform other duties. As if these were not problems enough, between 1980 and 1984 the number of employees operating the MIC unit over each three-shift period was substantially reduced: the main crew was halved to six members and the maintenance crew reduced from three to just one member, leaving two shifts without a maintenance supervisor. The reduced staff levels, combined with the high turnover of personnel, had a damaging effect on safety training: safe operating procedures were poorly understood and no training was given on the proper way to deal with emergencies.

In response to the safety audit, the plant management drew up a series of action plans with the purpose of rectifying the deficiencies. These were sent to the US parent company in order to indicate the progress made in improving operations. Some progress was indeed made in remedying the shortcomings. What was not addressed, however, were several design features of the MIC-manufacturing unit which, in combination with a high level of manual operation, substantially increased the risk of a serious accident. First, and most important, was the decision to provide for bulk storage of MIC rather than opting for either drum storage or a closed-cycle production process which did not require the storage of MIC. Second was the failure to ensure that the scrubber could neutralize not only escaping gases but also a mixture of gases and liquids. The significance of these two features was to be fully appreciated only when it was too late.

A further design error was made in 1983 in an attempt to simplify the maintenance of the unit. MIC was transferred from the storage tanks to the carbaryl production unit by pressurizing the respective tank with nitrogen; excess nitrogen was then vented from the system through a process pipe to the scrubber. Each storage tank was protected by a rupture disc; if pressure in the tank rose to a dangerous level, the disc would rupture and allow the MIC to be vented through a header pipe to the scrubber. However, in 1983 it was decided to connect the process pipe to the header pipe by a jumper line in order to facilitate the periodic removal of "trimer" (a residue which formed inside the pipes due to the reaction of MIC with small quantitites of water). This was done by flushing out the system with water, an operation which clearly had to be performed with great care so as to ensure that no water came into contact with the MIC.

One further failing was neither identified nor rectified: no emergency plan had been drawn up setting out the appropriate action to be taken should a major accident occur. Despite the known hazards posed by the plant's processes and the documented shortcomings in operating procedures, it is a remarkable and tragic fact that both local management and Union Carbide's US headquarters failed to ask themselves the basic question, "What if . . . ?"

THE ACCIDENT

With the benefit of hindsight it is easy to see that the conditions for disaster had been laid in Bhopal: an exceptionally hazardous process was in operation; the installation was technically flawed both in design and as a result of poor maintenance; operating personnel were inadequately trained and too few; plant procedures were sloppy; no emergency plan existed; regulatory control was ineffective; and densely populated shantytowns had grown up in close proximity to the site. The Union Carbide plant was an accident waiting to happen.

When it came, the accident was the inevitable result of a series of erroneous actions. The first of these mistakes took place in June 1984 when the refrigeration system for the MIC storage tanks was shut down to allow the coolant to be drained off and used elsewhere in the plant. As the temperature of the MIC in storage was bound to rise – it eventually stabilized at 15–20°C – the temperature alarm also had to be attended to. But instead of being reset to a higher temperature, the alarm was disconnected, with the result that no automatic warning of sudden heating could be given.

On 7 October, during the production of what was to be the last batch of MIC, the refining still operated at a higher temperature than was specified. This resulted in an MIC mixture with an abnormally high concentration of chloroform. However, the mixture was stored in tank 610 rather than in the off-specification tank 619. Two weeks later, on 21 October, it was noted that the nitrogen pressure in tank 610 had dropped from the usual 1.4 kg/cm^2 to 0.14 kg/cm^2, but the malfunction was not rectified. After the completion of the production cycle, on 23 October, the MIC unit was closed down for maintenance. At the same time the flare stack was disconnected in order to allow a section of corroded pipe to be replaced and the vent-gas scrubber was switched to stand-by, halting the automatic circulation of caustic soda through the stack. Five weeks later, on 30 November, an attempt was made to pressurize tank 610 but this was unsuccessful. At that time the tank was filled with 41 tonnes of MIC.

On Sunday, 2 December 1984, the production superintendent on the second shift instructed the MIC-unit supervisor to flush out part of the circuit to remove the trimer from the pipe walls. In order to ensure that no water came into contact with the MIC, it was standard practice to block off the pipes to be flushed out from the rest of the system. This was done with slip blinds which were inserted by the maintenance department. In this instance, however, the slip blinds had not been fitted; the operating personnel later claimed that no employee had been charged with this task since the position of maintenance supervisor on the second shift had been eliminated a few days previously.

Compounding these failures were two further defects in the MIC unit. First, some of the bleeder pipes – designed to draw off water at crucial points in the pipe runs – were blocked, allowing excess water to accumulate in the system. Second, several valves in the system were defective and had not been repaired. As a result water could escape from the circuit into the header pipe; from there the water could flow through the jumper line into the process pipe and down to the blow-down valve which controlled the pressurized nitrogen in tank 610. Whether the blow-down valve was sealed is not clear. What is clear is that attempts to pressurize tank 610 over the previous two weeks had failed, and it is quite possible that the blow-down valve was the cause of this failure, either because it was defective or because it had inadvertently been left open.

Shortly after 21.30, when the flushing operation had started, the operator observed that no water was coming out of the bleeder lines and he accordingly shut off the flow of water into the system. However, the MIC-unit supervisor instructed him to continue the flushing operation. At this time the pressure in tank 610 was seen to be 0.14 kg/cm^2, the same value as for the previous few days.

The third shift took over the operation of the MIC unit at 22.45. At 23.00 the new control-room operator observed a pressure reading in tank 610 of 0.70 kg/m^2. There is some uncertainty as to whether the new operator had been informed of the earlier reading of 0.14 kg/m^2. However, a pressure of 0.70 kg/m^2 was well within the normal operating limits of the unit, and since temperatures inside the tank were not logged the operator had no means of knowing if anything was amiss. Then, at 23.30, a field operator noticed that the unit was leaking near the scrubber; a mixture of water and MIC was escaping from a branch in the header pipe. Personnel were called to inspect the leak and the control room was then informed of the problem.

Just after midnight, at 00.15, the control-room operator checked the pressure reading in tank 610 again and observed that it had increased to 2.1 kg/cm^2 and was rising rapidly. As he watched, the reading reached 3.9 kg/cm^2 and then went off the scale. The operator called the MIC-unit supervisor and ran outside to inspect the tank. A screeching noise from the safety valve indicated that the rupture disc had blown; heat was radiating from the unit and rumbling noises could be heard from inside the tank.

Returning to the control room, the operator switched on the scrubber in order to neutralize any escaping gases. But the scrubber apparently failed to operate – the flow meter did not register that the caustic soda was circulating. At the same time workers outside saw a cloud of gas burst out of the stack.

At about 01.00 the toxic gas alarm was sounded for a few minutes to warn the local community of the emergency. The firewater sprays were switched on and directed at the storage tanks and the stack, although the water did not reach high enough to blanket the cloud. An attempt was also made to turn on the unit's refrigeration system, but this was futile as the coolant had been drained off six months previously. There was little for the emergency crews to do; only when the reaction died down at around 02.30 did the safety valve reseat, once more sealing the tank. But by then the release had become the worst industrial accident in history; 35 tonnes of MIC had escaped into the atmosphere in the middle of a densely populated urban area.

THE CONSEQUENCES

At the time of the release, weather conditions were stable, with a light north-northwesterly breeze blowing and an ambient temperature of about 15°C. The breeze carried the cloud of gas along the main railway line that ran past the Union Carbide plant. Being heavier than air, the gas remained close to the ground and dissipated very slowly. The area most seriously affected was a path about 3 km wide and 10 km long, although effects were felt as far as 80 km from Bhopal.

It is estimated that some 200 000 people attempted to flee from the gas. About 500 were to die that night in Bhopal, suffocating after being enveloped by the cloud: nobody knew that the most effective protection was to simply lie down with a wet cloth covering the face.

At dawn, the southeast quadrant of the city resembled a battlefield. Corpses lay everywhere, and thousands of injured inhabitants beseiged the four hospitals. But with only 1800 beds and 300 doctors available, the task was hopeless. Further, nobody knew how to treat the victims – not least because it was not known what toxic gases had escaped.

Indeed, the chemical make-up of the cloud was never established with certitude. Without doubt it contained MIC contaminated with chloroform, but other reaction products also escaped. One of these compounds may have been hydrogen cyanide, and some autopsies indicated symptoms which matched those associated with cyanide poisoning. This possibility later generated a heated local controversy, for if the

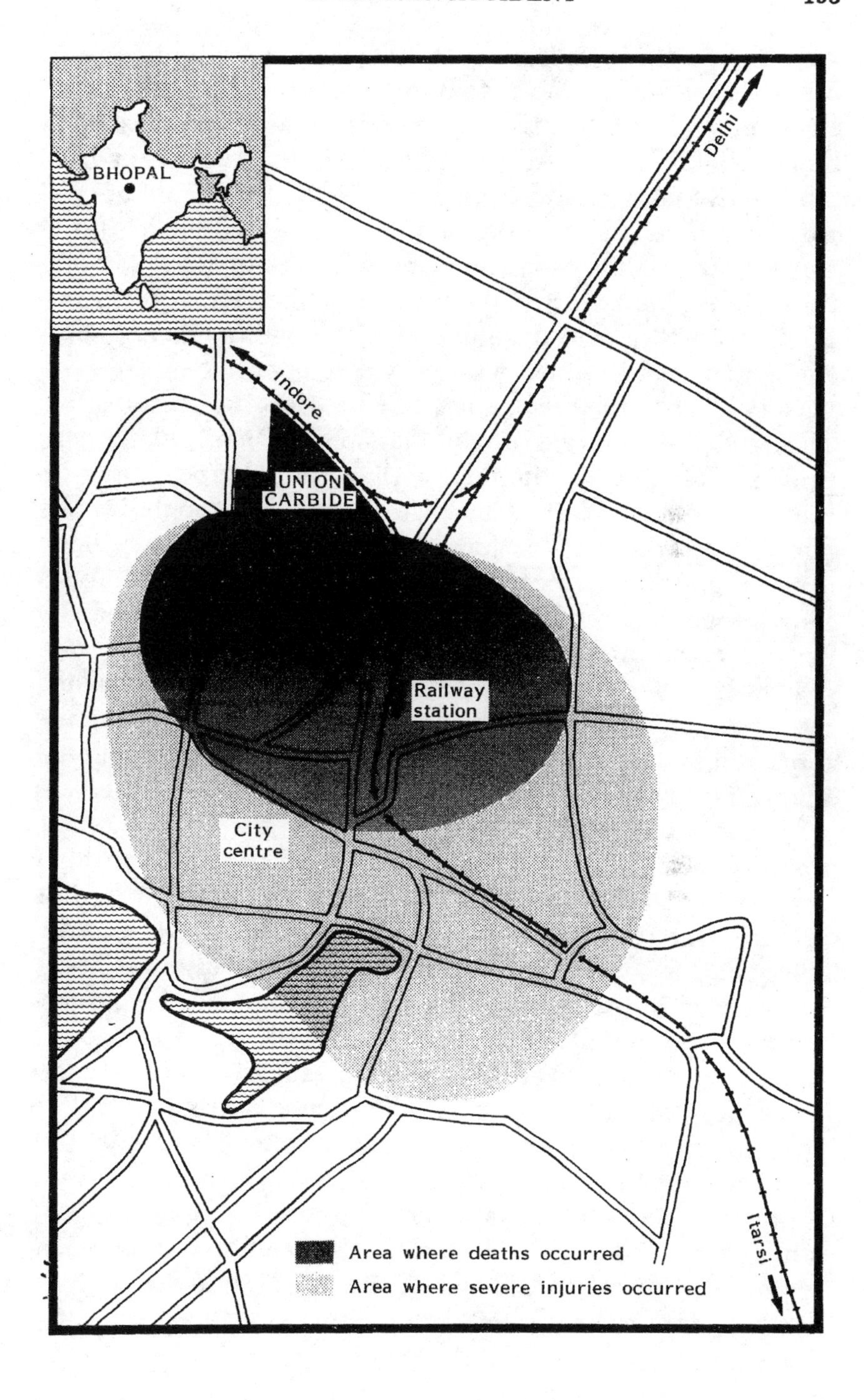
BHOPAL
Delhi
Indore
UNION CARBIDE
Railway station
City centre
Itarsi
Area where deaths occurred
Area where severe injuries occurred

victims had suffered from cyanide poisoning, administering the well-known antidote sodium thiosulphate would have saved many lives. But this was not done, and Union Carbide has consistently denied that cyanide could have escaped.

With the local medical services completely unable to cope with the disaster, the death toll grew rapidly. Within a week nearly 3000 people had died and a further 250 000 inhabitants were suffering from the effects of exposure to the gas. In this period immediately following the accident, the major relief effort was provided spontaneously by a variety of organizations, such as charities and social clubs. They collected food, arranged transport to medical centres, provided temporary shelter for the orphans and helped to bury or cremate the victims. It took longer to mobilize the government relief operation – which was hardly surprising given the scale of assistance which was necessary – and a new Department of Relief and Rehabilitation was specially set up for the purpose.

It will never be known exactly how many people died as a result of exposure to the gas. Because of the chaotic situation which prevailed immediately after the accident and the overburdening of the local medical resources, it was impossible to maintain an accurate record of the victims. Families often arranged funerals for their own dead without registering the victims with the local authorities, hospitals released many of the bodies to relatives for private burial or cremation, and there was no way of knowing how many people died in the areas to the south of Bhopal. The government's own estimate of the final death toll was 3135, though some officials admit that this was almost certainly an underestimate. Estimates based on eye-witness accounts and on circumstantial evidence, such as the number of shrouds sold, suggest that between 8000 and 10 000 people died. Many times that number suffered serious health damage: nearly a quarter of the established pregnancies ended in abortion, stillbirth or premature death of the infant, and many thousands of inhabitants were permanently disabled, the most serious ailments involving the respiratory tract, such as bronchitis and fibrosis. Subsequent surveys indicated that, of the 250 000 affected by the disaster, about 60 000 were severely disabled and a further 40 000 partly disabled.

The economic and social ramifications were even more widespread. The Union Carbide plant closed down, leaving over a thousand workers unemployed. Support industries also suffered greatly, and tax revenues to the state and local governments fell accordingly. Damage to local business has been estimated at anywhere between $8 million and $65 million. Families had to cope not only with bereavement but also with the loss of earning power and with the problems of carrying out gruelling domestic chores, such as gathering firewood and fetching water. Many bereaved women, already disadvantaged through their sex or their caste, became prey to exploitation, and many families found themselves being driven further and further into debt.

THE RETRIBUTION

A well-publicized event in the aftermath of the disaster was the arrival in Bhopal of several lawyers from the US. These lawyers, specialists in personal injury claims, made contact with local law firms and the victims of the accident with the aim of bringing multi-million-dollar lawsuits in the US against Union Carbide. The result of the efforts of the lawyers was a burgeoning number of lawsuits, both in the US and in Bhopal itself. There was, however, little coordination in this activity. Moreover, it soon became apparent that there were several parties who might be considered to have been negligent: not only Union Carbide, but also the state and national governments for authorizing the operation of a hazardous plant, for legalizing the slum colonies in close proximity to the plant, for failing to ensure that the plant was properly maintained and that safety procedures were adhered to, and for omitting to draw up an emergency plan which would take effect in the event of a serious accident at the plant.

Within a matter of weeks personal injury suits were brought against Union Carbide in several different state and federal courts in the US; many more were brought against the Indian division of the company in Bhopal. These included a claim for $1 billion filed in Bhopal against both the Indian division of Union Carbide and its parent company, and a suit filed in the Supreme Court of India against Union Carbide (India) Ltd, the government of India and the state government of Madya Pradesh.

In March 1985, however, three months after the accident, the Indian government passed the Bhopal Gas Leak Disaster (Processing of Claims) Ordinance. This legislation vested the government with the power to represent all the victims of the disaster and to coordinate all aspects of the claims. On 8 April 1985, the Indian government filed a claim in the US against Union Carbide on behalf of the victims, covering both personal injury and property damage, and all the actions brought in the US, totalling over $100 billion; these were then consolidated in the US District Court in Manhattan.

It was to take four years to reach a settlement. And as is often the case in major personal injury lawsuits, the court proceedings were paralleled by negotiations between the plaintiffs and the defendants on a mutually acceptable out-of-court settlement. Union Carbide's first offer was made in August 1985 and amounted to $200 million to be paid over 30 years, but the Indian government rejected the offer. In March 1986, shortly before the court was due to pronounce on whether the case should be heard in the US or India – with all that that implied for the size of any settlement, US courts traditionally fixing extremely high monetary damages – Union Carbide raised the offer to $350 million. Again the Indian government rejected the amount as too low.

In the event the District Court decided that the case

should properly be heard in an Indian court of law. The suit was duly brought in Bhopal in September 1986, with the Indian government claiming damages from Union Carbide for personal injury on behalf of the victims and for environmental damage totalling $3.12 billion. The case was to go to the Court of Appeal and then to the Supreme Court in New Delhi for a judgment on the settlement. Judge Pathak, fully aware that the best interests of the victims lay in a speedy and final resolution of the claim, went to some lengths to force the case to a conclusion, and on 14 February 1989 the court pronounced judgment on the appropriate level of damages: $470 million to be paid to the Indian government by 31 March. However, in the interests of expediting the resolution of the issue, the verdict specifically excluded further claims against the company. This aspect of the verdict was sufficiently important to Union Carbide for it to accept the ruling immediately. But the victims – and subsequently the Indian government – saw the settlement as inadequate, and attempts to overturn the ruling were initiated. For the lawyers, the judicial trial is likely to last for some years; for the 100 000 victims still being treated for their injuries the personal trial will last far longer.

CHAPTER SEVEN

A BRIDGE TOO LOW

The gateway to the Magra valley

Anyone who has visited the northwest coast of Italy, where the Apuan Alps and the Apennines drop down to meet the Mediterranean, knows it to be a region of exceptional beauty. The rugged peaks, reaching up to 2000 m provide a dramatic backdrop to the richly wooded slopes skirting the eastern shoreline of the Gulf of Genoa. Opportunities for cruising along the coast, exploring the varied landscape or just relaxing at the water's edge combine with an unusually amenable climate to make the area highly attractive to tourists.

It is on this coastline, just to the south of the historic port of La Spézia, that the river Magra discharges its clear, fast-flowing waters into the Tirreno Sea. And it was here, in 1960, that the Columbiera Bridge was built, extending the coast road over the river to the picturesque village of Amèglia and on to Romito. Constructed to improve communications and stimulate the local economy, the bridge saves those wishing to cross the estuary a 5 km journey inland to Sarzana. It is ironic, then, that it was this bridge that ultimately served to prevent the industrial development of the Magra valley. And it did so in just about the most dramatic way imaginable – by blocking access to the sea for six new warships, built at a cost of $500 million in a shipyard just upstream of the bridge.

The question raised by the issue is an evergreen dilemma:

how do we value a natural resource? It is also a question with a special significance for environmental management, for if we cannot devise a common calculus by which to compare the benefits of industrial development with the costs of the resulting environmental degradation, how can we hope to make proper decisions regarding the propriety of such actions? Unfortunately, the task of devising a generally applicable method for making such trade-offs has defied the best efforts of economists for generations. We are therefore confronted with the awkward problem that the benefits of economic activities which cause environmental impacts can be expressed in hard dollars but the environmental damage cannot. It is hardly a fair contest. In this instance, the dilemma was posed in a very straightforward form: did the preservation of the Magra valley justify writing off $500 million-worth of newly built ships? Even so, the answer proved to be far from simple.

GRAND CENTRAL

The European environment as we know it today is largely a product of glaciation. For more than 2 million years the great ice sheets ebbed and flowed, transforming both the geomorphology and the ecology of the continent. The distribution of plant and animal species varied greatly during this period as their favoured habitats shifted to higher or lower latitudes in phase with the advance and retreat of the glaciers. In southern Europe the extent of these migrations was hindered by two great natural barriers: the Pyrenees and the Alps. These two mountain chains are 250 km apart, and it was into this narrow gap that a multitude of plant and animal species were repeatedly funnelled.

Guarding the southeastern approaches to this Pleistocene "Grand Central" was the Magra valley. Fed by a river system covering some 1700 km^2 of the southwestern slopes of the Apennines, the valley is divided for much of its length

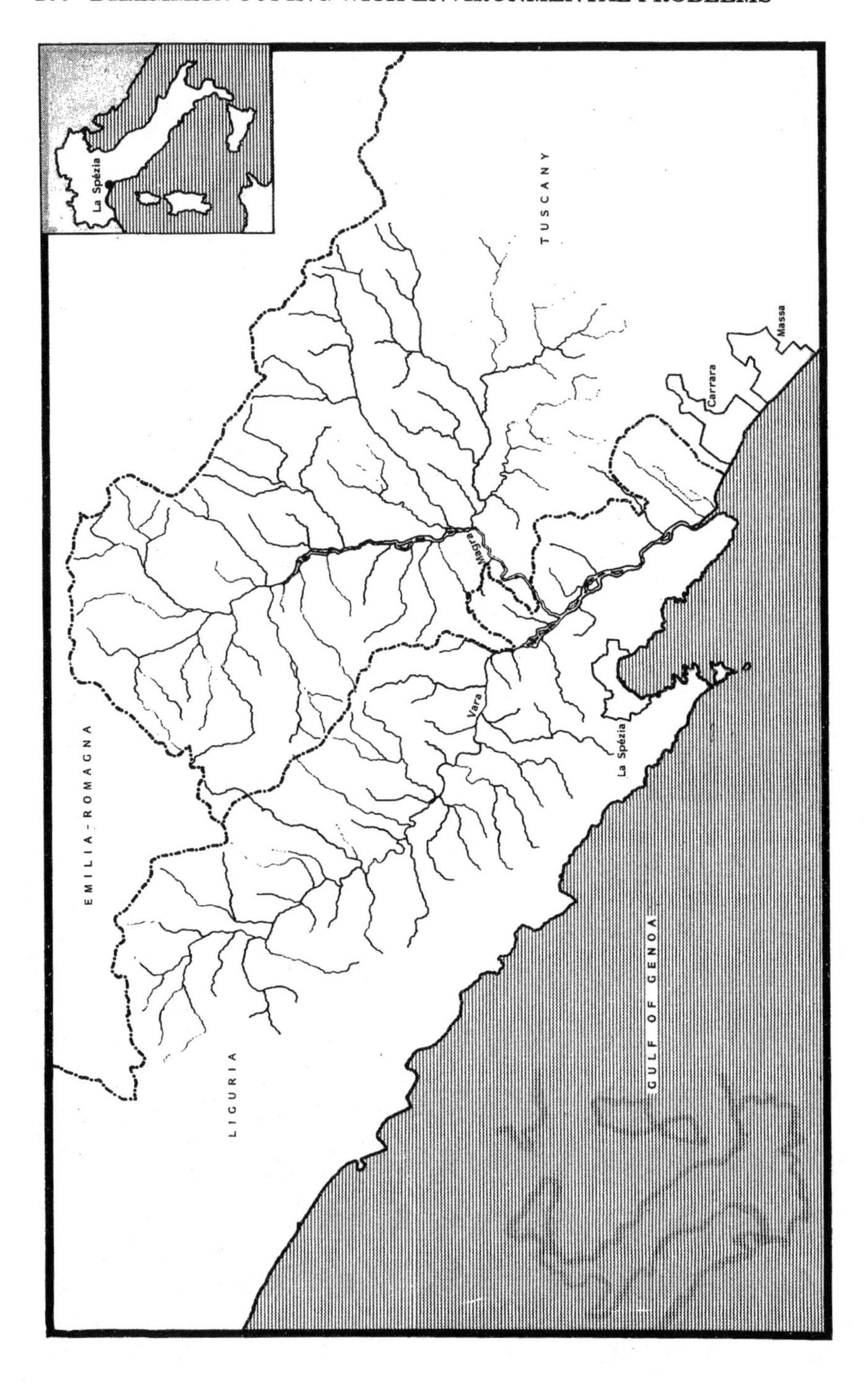
La Spézia
TUSCANY
Massa
Carrara
Magra
Vara
La Spézia
EMILIA-ROMAGNA
LIGURIA
GULF OF GENOA

into two separate drainage basins: that of the Magra itself and, almost as extensive, that of the river Vara. These catchment areas converge at the village of Bottagna, where the river broadens out into a braided stream for the final 15 km of its journey to the Tirreno Sea, depositing much of the sand and gravel transported down from the mountain gullies onto the gently sloping Piedmont valley floor.

In many ways the Magra is not untypical of river systems fed from steep mountain formations, in which the barren slopes shed large quantities of course debris into the fast-flowing streams and the lower valleys are subject to aggradation and periodic flooding. But it is certainly exceptional in the richness of its biological heritage, a heritage still largely intact in the wide range of microclimates to be found in the valley. Various combinations of temperature, rainfall, exposure and hydraulic regimes offer a multitude of ecological niches, both terrestrial and aquatic. The result is a river basin unique in Italy for the diversity of its ecology. That these exceptional conditions survived into the late twentieth century is remarkable and largely due to the fact that the valley was, for many years, spared the disruption caused by some of the more intrusive human activities. But the changes, when they did come, were profound.

TWO ROADS

For centuries the economy of the Magra basin was dependent on horticulture in the lower river valleys and fishing in the shallow coastal waters. In recent years this traditional way of life has been precipitately disrupted by a precocious intruder – tourism. An important stimulus for this development was the proximity of the Forte dei Marmi/Viareggio complex a few kilometres south along the coast, one of the most popular resorts in Italy. Its impact on the way of life in the valley proved to be momentous: within a few years the local fishing industry had been almost entirely displaced by recreational boating enterprises.

Then, in the 1960s, two new autostradas were constructed: the A12 coastal route from Genoa to Livorno and, following the river Magra down the slopes of the Apennines, the A15 linking Parma with La Spézia. These roads encouraged the growth in tourism still further. They also had two other effects. The first was to improve communications with the port of La Spézia, thereby stimulating the development of an already buoyant shipbuilding industry in the town. But La Spézia, a natural harbour in the shadow of the Ligurian Apennines, was by then highly congested with only a limited area of land remaining to be developed. Industry therefore looked to the Magra estuary, just 12 km along the coast, as a possible location for further expansion, and plans were even mooted for dredging the river to a depth of 7 m in order to provide a shipping channel into Sarzana.

The second impact of the new roads was on the river valley itself. Laying an autostrada along the rugged landscape of the Magra valley required huge quantities of sand and gravel. Conveniently for the engineers, these materials were in abundant supply in the valley itself, and the river bed was heavily exploited for this purpose. The result of these excavations was to substantially reduce the extensive areas of shallow water along the margins of the river which had provided a valuable habitat for many aquatic species. Not only that, but the changes to the river profile had a serious effect on stream flow, with the result that erosion increased noticeably along the river banks. Removal of sand and gravel from the river bed, not only for road building but also for housing and other construction work, had in fact been going on illegally and on a large scale for a long time. Lax enforcement – a recurrent theme in this tale – allowed the practice to continue, even when the consequences of uncontrolled extraction became clear. The most dramatic illustration of the problem was provided in 1978 when the old stone bridge at Romito suddenly collapsed after its foundations had been undermined.

And it was not only the river that suffered. Some of the deeper excavations in the lower reaches of the Magra penetrated the clay stratum underlying the gravel bed, allowing the saline water of the estuary to permeate the deeper-lying aquifers. This posed a serious threat to the region, for some 300 000 people were dependent on these aquifers for their drinking water. Yet, with no remedial measures being taken, the contamination of the groundwater became more and more extensive, so that by 1986 salt concentrations of 4 g/l had been found 4 km inland from the sea.[1] As if this was not serious enough, some of the deep ponds left by the excavations were being used for the illegal dumping of industrial wastes, with the result that toxic compounds were beginning to be detected in the groundwater.

A SHIPYARD

It was in these circumstances that a new boat-building company, Inter-International Marine of Italy SpA, later to become simply Intermarine SpA, started looking for a suitable site on which to construct a shipyard. Registered in La Spézia in 1970 by Robert Sutz, Arthur Weber (both Swiss nationals) and Rocco Canelli (Italian and designated Administrative Director), the company was launched to capitalize on the growth in sales of pleasure boats from Italian yards. The founders were aiming at the market in luxury, high-performance vessels up to a length of 15 m. The initial intention of establishing the shipyard in La Spézia was soon seen to be impractical due to the scarcity of suitable sites, and a search for alternative locations was begun. It quickly became apparent that a site on the east bank of the river Magra offered both sufficient space – a total of

[1]This compares with the generally accepted quality standard for drinking water of 200 mg chloride/l, equivalent to 330 mg salt/l

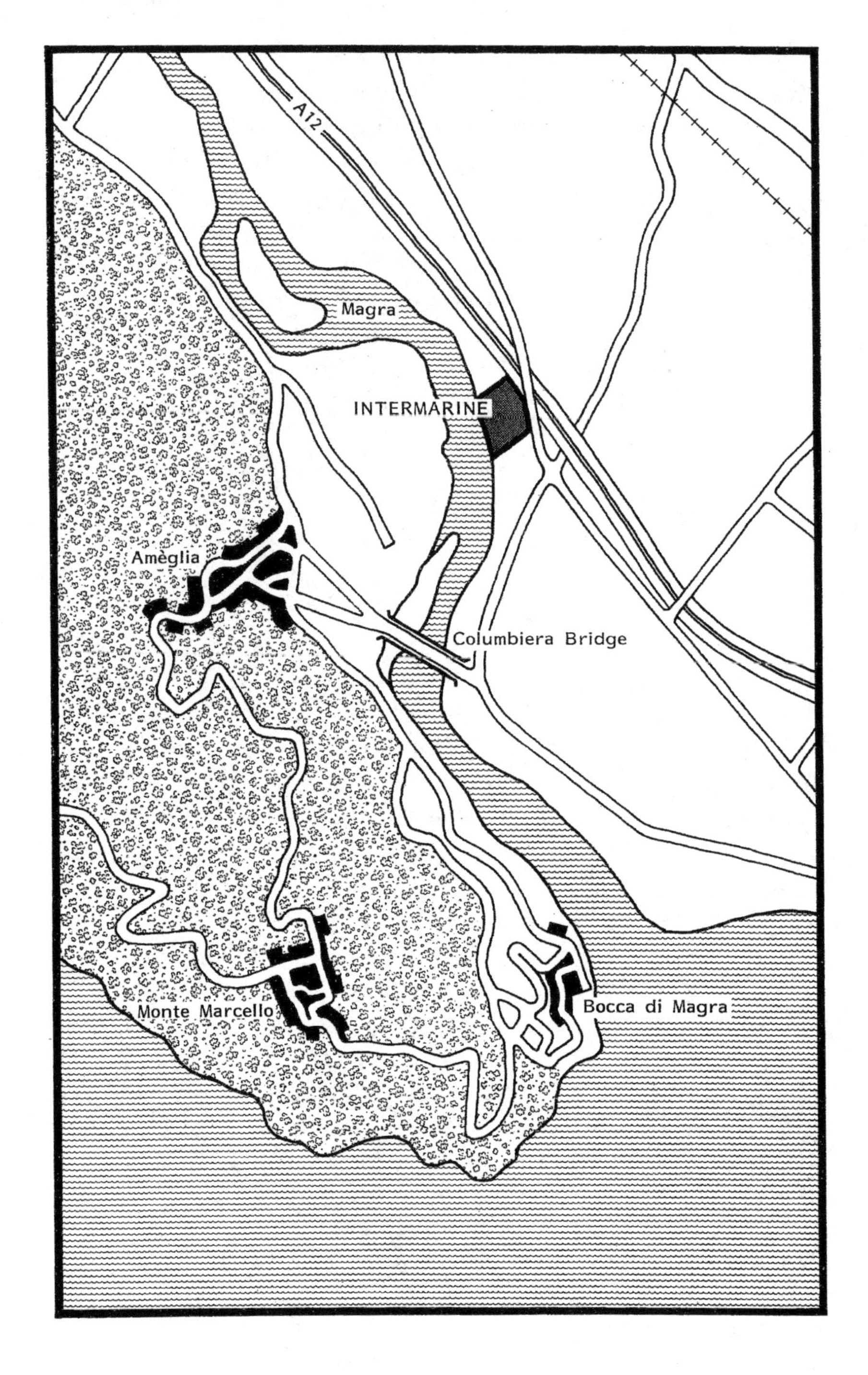
A12
Magra
INTERMARINE
Ameglia
Columbiera Bridge
Monte Marcello
Bocca di Magra

74 000 m^2 – and easy access to the sea. It was, to be sure, 500 m upstream of the Columbiera Bridge, but the headroom of 6 m under the spans was easily sufficient to allow passage for the relatively small boats which were to be produced by Intermarine. In fact no problems were foreseen for bigger vessels either as the mayor of Sarzana, Anelito Barontini, keen to attract new enterprises to his commune, indicated to Canelli that it would be a straightforward matter to modify the bridge should it ever prove to be an obstacle to larger ships.

Armed with this assurance, Intermarine went ahead with the purchase in November 1970, and over a period of two years created a modern, well-equipped yard on the site. Intermarine was by no means the first industrial undertaking to locate on this stretch of the river, being preceded by the shipyards Euromare and Metalcost and several small boatyards. It was, nevertheless, the largest of the enterprises, and it was therefore notable that the purchase of the site and the preparations for construction went ahead before a building permit had been granted from the commune of Sarzana. This was a mandatory procedure under the Town Planning Act, intended to ensure that any new development complied with local planning objectives and public health regulations. It should be noted that not only did construction proceed without a permit, but that the land in question was zoned for agricultural and tourist use under the local development plan.

As events would repeatedly demonstrate, however, Rocco Canelli was an exceptionally persuasive character. In this instance, the outcome of his discussions in Sarzana was a permit to construct a shipyard on the site, signed by Mayor Barontini on 9 December 1970. The fact that the development plan was not amended – thereby making the permit invalid – only became an issue some years later, and even Intermarine eventually conceded that the permit "probably" did not comply with planning regulations. There was, moreover, a further irregularity which was to presage the subsequent traumatic events. It later came to light that not one but two permits were issued to Intermarine simultaneously. They differed in a single respect only: one authorized the construction of a shipyard specifically for the manufacture of "pleasure boats" ("*imbarcazioni da diporto*"), the other simply for "boats" ("*imbarcazioni*"). No explanation has ever emerged for this curious anomaly, either from the commune or from the company. But it was precisely a dispute as to the type of vessel assembled by the company which was to spark a controversy that ultimately went all the way to the Italian cabinet.

THE CALM BEFORE THE STORM

Despite the dubious legitimacy of the new undertaking, few signs were visible of any animosity towards Intermarine within the local community. Taking advantage of this accommodating atmosphere, the company was able to concentrate its efforts, in the early years, on developing and marketing its pleasure boats and on improving the facilities in the shipyard – somewhat over enthusiastically as it transpired, for while the original building permit authorized a maximum enclosed area of 760 m^2, faced in aluminium, Intermarine proceeded to erect a construction hall with fibrous cement facades enclosing an area of 13 700 m^2.

Central to Intermarine's commercial strategy was the application of fibreglass technology. The great advantage of using this material in boat construction is the scope which it offers for producing complex, lightweight structures in one-piece mouldings. It was largely at the initiative of Intermarine's South African technical director, Michael Trimming, that the technique was exploited in the company's range of pleasure boats. Indeed, Trimming saw a far greater potential for the material and drew up various designs for naval patrol craft up to 27 m in length. A moderate degree of interest was shown in these vessels, and

Intermarine produced and sold several examples for military purposes.

This first venture into the naval market clearly suggested great possibilities to Canelli and Trimming, and an opportunity to exploit the potential was not long in arriving. By the early 1970s it had become apparent to NATO, the Western defence alliance, that the minesweepers then in service were rapidly being rendered obsolete by recent advances in mine technology. A specification was accordingly drawn up for a new type of ship, a so-called "mine countermeasures vessel", or simply minehunter, which would go into service with the navies of the various NATO countries. This specification covered the vital operational parameters of the vessel, and each member country was at liberty to place an order with a suitable shipyard provided the design of the minehunter corresponded with the NATO requirements. Italy, as a member of the alliance, duly adopted the new specification, and in 1975 parliament authorized the Ministry of Defence to order a total of 10 new minehunters.

The specific problem confronting NATO was the development of "intelligent" mines. Unlike traditional mines, which floated on or just below the surface of the water and relied on chance to be struck by a ship, intelligent mines lie on the sea or river bed and are programmed to respond to vessels passing near them. They may be activated in several ways: by registering the magnetic field of the ship, the noise of the engine or the pressure wave generated by the hull as it moves through the water. In fact mines are now capable of recognizing the magnetic signatures of different types of ship – a cruiser, a tug or a fishing boat, for example – and responding accordingly. This capability places considerable demands on the performance of minehunters. They must be silent, flat-bottomed so as to reduce the magnitude of the pressure wave, and strong enough to withstand an explosion, should this prove to be necessary.

Traditionally minesweepers have employed a wooden-hulled construction to overcome the problem of magnetism, and it was here that Intermarine saw an opportunity for an important advance and the chance to gain a competitive edge over other shipyards. It seemed to the company that fibreglass would be an ideal material for the purpose as it is non-magnetic. Moreover, it could absorb the force of an explosion by flexing and would allow the hull to be moulded in one piece, thereby eliminating the joints which had proved to be the weakest part of the wooden vessels. Intermarine's design team accordingly set to work and produced a basic design for technical evaluation by the navy's engineers. This proposal featured the revolutionary idea of a one-piece monocoque hull, that is to say, the skin of the vessel was built up from multiple layers of fibreglass to produce a load-bearing hull. Designed without any internal bracing, the flexibility of the hull was calculated to absorb not only explosions but also the vibrations of the ship's components, thereby reducing noise during operations.

The preliminary assessment by the navy was generally positive, but the novel nature of the design made it essential that a series of tests be carried out in order to demonstrate the integrity of the concept. It was here that the first signs of irregularities in the relationship between Intermarine and the Ministry of Defence appeared. Under normal circumstances, the responsibility for demonstrating the worth of a design proposal when tendering for a navy contract lies entirely with the manufacturer: the company will be required to finance all research and development, including the construction of any prototypes. In this case, however, Intermarine's strained financial position did not permit it to make the substantial investment necessary to develop such a revolutionary design. It was necessary, for example, to supply the navy with a complete centre-section of the minehunter so that its resistance to explosions could be tested. The production of this component required the manufacture of the moulds for laying up the fibreglass hull, and this was a costly undertaking. In the event, much of this research and development was financed by the navy itself through a payment of L 815 million ($2 million). The reasons for this exceptional treatment are not clear. Given that it was Intermarine's first attempt to win a major defence contract, that the vessel was of a type with which it had no previous experience and that the company's financial viability left something to be desired, one would have expected the ministry to act with some caution. The preferential treatment accorded Intermarine created suspicions amongst its competitors and in the press that the dealings between the company and the navy were not entirely above board.

This scepticism was exacerbated by a number of other curious factors relating to Intermarine's involvement. Over 90 per cent of the company's capital was of Swiss origin, having been advanced by Simonin AG of Zurich and Verkehrs Aktiengesellschaft of Lausanne. This in itself is not unusual, but the practice in Italy, as in many other countries, is that defence contracts should be awarded, where feasible, to Italian companies. Moreover, under the navy's system of classifying its suppliers, Intermarine fell into Class 9 – signifying that its capacity was such as to justify orders amounting to a maximum of L 6 billion ($15 million) a year. Yet the minehunter contract was to involve average annual expenditures of over L 9 billion ($23 million). The fact that Intermarine was not equipped to build the minehunters at the time of tendering was also somewhat unusual. Its construction hall, for example, was too low to accommodate the vessels and there was no basin in the yard where outfitting could be completed.

Three of Intermarine's competitors made more specific allegations. In effect, the Baglietto, Picchotti and Italcraft shipyards maintained that the navy had decided to give the order to Intermarine from the very beginning, pointing to a number of irregularities. The official invitation to tender for the contract was sent out in April 1976 and allowed just 60 days for the submission of a proposal. This was an exceptionally short period in which to draw up a design, and the three companies, working together as a consortium, succeeded in gaining an extra 15 days for the completion of their proposal. However, they claimed that Intermarine had already been working on the project for over a year, liaising with the navy's technicians for most of that time. Indeed, it was even said that one of these technicians had actually worked on Intermarine's premises for 10 months. The competitors heard, unofficially, in June 1977 that the order was to go to Intermarine, but on examining Intermarine's design they found it to depart from the original specification to which they had worked. In particular, they pointed to the fact that Intermarine's proposal made allowance for only one engine instead of two. Nevertheless, these allegations failed to persuade the Minister of Defence, Attilo Ruffini, to review the contract. An internal briefing paper, prepared by the Secretary-General of the ministry and dated 21 March 1978, advised the minister that Intermarine's design was an important development for the Italian shipbuilding industry. Moreover, a review of the order would set the minehunter programme back by 18 months and could have serious legal implications.

But there was an even greater and more obvious encumbrance to Intermarine's attempt to win the minehunter order. The company's design was for a 50 m, 500 tonne vessel with a superstructure rising 8.5 m above the waterline. The Columbiera Bridge, on the other hand, afforded a clearance beneath its spans of just 6 m. Even if Intermarine was awarded the contract, getting the ships to the sea was clearly going to be no easy matter.

The implications of this difficulty were not lost on the company. One thing was certain: a solution to the problem had to be found if Intermarine was to have any chance of getting the order. Rocco Canelli therefore approached the national road authority, Azienda Nazionale Autonoma della Strade (ANAS), and broached the question of modifying the bridge so as to allow passage for the minehunters. The first formal record of this request is to be found in a letter dated

16 July 1976, from Intermarine to the regional office of ANAS in Genoa. In the letter, Intermarine set out its plans to build larger vessels, pointed to the fact that the Columbiera Bridge provided just 6 m of headroom, and requested permission to replace one of the fixed spans with an opening section. The letter added that the forthcoming expansion of the shipyard offered the prospect of more than 300 new jobs and that opening the bridge would also be to the advantage of other industries in the area, thereby stimulating further economic development. On the other hand, it warned, failure to realize the plans would lead to the laying off of some of Intermarine's existing employees. The company's ensuing discussions with the road authority were undertaken with the regional director of ANAS, Ernesto de Bernardis. Although the request to modify a publicly owned road bridge for the sole purpose of enhancing the commercial interests of a private company might seem rather unusual, Rocco Canelli's arguments were again highly convincing. De Bernardis indicated that permission would be granted and, given this assurance, Intermarine submitted a formal request to ANAS on 16 December 1976.

Under normal circumstances, the lengthy bureaucratic procedure of approving such an application could be expected to take months or even years. In this instance, however, an agreement between ANAS and Intermarine was drawn up by De Bernardis and Canelli in the space of just eight days, to be signed on 24 December just as everything was closing down for Christmas. The agreement authorized Intermarine to modify the Columbiera Bridge so as to allow the passage of its ships. Both construction and maintenance costs were to be borne by Intermarine, although in return the company was to have sole rights to the operation of the new opening section, which could be opened once a week for a maximum of 60 minutes. The agreement was to be valid for ten years and extended for a further ten years in the absence of written notification to the contrary. A formal endorsement of the agreement was signed by De Bernardis on 11 January 1977.

It should be pointed out that the unprecedented speed with which ANAS dealt with Intermarine's application had a simple explanation: in his haste to approve the modifications, De Bernardis unfortunately omitted to notify the various bodies which he was legally obliged to consult. These included the communes of Amèglia and Sarzana, the Ligurian regional public works authority, the La Spézia harbour authority and the gas, electric, water and telephone utilities which routed their mains and cables across the Columbiera Bridge. His enthusiasm in this respect was subsequently to have serious consequences for both him and Intermarine.

As far as Rocco Canelli was concerned, however, the problem posed by the bridge was resolved; Intermarine could now fully commit itself to securing the minehunter order. This it did with some gusto; by the summer of 1977 things were looking good for the company, and in December a navy representative let it be known that the contract would be concluded with Intermarine by the end of the year. In fact it took slightly longer, but on 7 January 1978 the contract was duly signed: Intermarine was to supply the Ministry of Defence with four minehunters, the first to be delivered in La Spézia by 13 May 1981, the second by 30 August 1983, the third by 25 June 1984 and the fourth by 28 February 1985. The total contract sum amounted to L 64 billion ($160 million). Moreover, given the government decision to purchase up to ten such vessels, there was obviously the prospect of a further order. Indeed, only four months later, on 20 April, the Ministry of Defence took out an option on two more minehunters. For a shipyard largely financed by foreign capital and, furthermore, suffering from serious financial difficulties, sited in contravention of the local development plan, constructed at variance with its building permit, separated from the sea by a low bridge, and with no previous experience of working on large vessels or major defence contracts, the future was looking very rosy indeed.

CLOUDS ON THE HORIZON

After eight years of tribulations, Intermarine must have thought that it was at last running before the wind. Everything seemed to be falling into place: the local administration had turned a blind eye to the unfortunate contraventions of its zoning and building regulations and accepted the fact of the shipyard's presence; the order for the minehunters at last promised a secure commercial future; and the regional head of the roads authority had been persuaded to approve modifications to the Columbiera Bridge. By now, however, people with a rather different interest in the Magra valley were putting two and two together and starting to become concerned at the results of their calculations. Amongst these onlookers was the environmental group Italia Nostra. Founded in 1955, Italia Nostra (literally, "Our Italy") had developed into an effective campaigning organization through its astute use of the law, often being granted legal standing in court actions as a representative of the collective public interest in environmental protection. As luck would have it, the president of the district branch in which the Magra estuary was sited was Dr Luigi Biso, a retired hospital director who lived in his family home a kilometre downstream from the Columbiera Bridge. Although he was amongst the first to recognize the threat of industrial development in the valley, Biso had originally pursued a moderate line, acknowledging the employment advantages offered by the new boatyards and hoping for sympathetic application of the zoning laws. But with the announcement of the minehunter order and the knowledge that Intermarine was to start producing far larger vessels, Biso began taking a much more active interest in the issue.

His first step was to write to the authorities concerned, requesting information on the reported developments. Admirals Torrisi (Chief of Staff) and Caporali (Director of the Coastguard) confirmed the order for the minehunters and explained that Intermarine had been granted all the necessary authorizations to carry out the work. A rather different story was told by the Ligurian Public Works Department. This regional authority denied being notified of any proposed modifications to the Columbiera Bridge. Even more worrying, the national office of ANAS knew nothing of the contract between its Genoa office and Intermarine. With the local press starting to scent a good story, letters began to circulate amongst the various authorities in an attempt to find out exactly what had happened. First, in October 1978, the commander of La Spézia harbour authority, Giuliano D'Este, wrote to ANAS, the head of the Ligurian Public Works Department, the Maritime Civil Engineering Department and the mayor of Amèglia, Franco Baldassari. In the circular letter he noted the press reports of a major change to the Columbiera Bridge, explained that this stretch of the river Magra fell within his jurisdiction and requested information on the proposals. On learning that an agreement had been signed permitting Intermarine to replace one of the bridge's fixed spans with a movable section, the commander promptly dispatched a telex to ANAS expressing his surprise, demanding more information and warning that he would prosecute the authority if work was begun without his consent.

The news also went down badly with the commune of Amèglia. Shortly after D'Este's inquiry, Mayor Baldassari wrote to the Ministry of Public Works in Rome and to the regional office of ANAS in Genoa pointing out that questions were being asked in the council, and requesting more information on the proposals. To Intermarine, it was now clear that a broader public needed to be persuaded of the propriety of its intentions; it accordingly decided, perhaps somewhat belatedly, to make a public announcement on the minehunter programme. This it did on 7 November 1978. A formal notification was sent to the mayor of Amèglia with copies to the mayor of Sarzana, Francesco Baudone, the president of the province of La Spézia, Ferdinando Pastine, the Industrialists Union of the province of La

Spézia, the Sarzana branch of the United Federation of Trade Unions and the local branch of Italia Nostra. The letter confirmed Intermarine's success in winning the order and set out its contractual obligation to supply the Ministry of Defence with four minehunters by 1985. It went on to explain that alterations to its construction facilities would be necessary, and that the workforce would have to be expanded from 150 employees to as many as 250. ANAS had already granted approval to modify the Columbiera Bridge, although it would only need to be opened a few times each year. Finally, Intermarine assured the mayor that the construction of the minehunters would not cause any pollution of the river Magra.

The first reactions to the announcement were enthusiastic. Both the Industrialists Union and the United Federation of Trade Unions saw Intermarine's plans as a positive development for the area. The possible environmental implications were not a cause for serious concern, the trade unions dismissed any threat to the Magra valley with the assertion that "an open bridge does not pollute the environment". But it was this very enthusiasm which carried the seeds of conflict. Indeed, it was the response of the Industrialists Union which sparked a controversy in the local community over the issue. In his letter to Intermarine of 14 December 1978, the president of the Union, G.B. Rosa, concluded that the opening of the bridge was in the general interest and, as far as the siting of other industrial undertakings was concerned, it could "arouse the interest of the marine engineering industry in particular". In short, he gave a clear indication that local industrial interests were actively considering further development along the Magra. Rosa distributed copies of this letter to the original recipients of Intermarine's notification. As these included Italia Nostra amongst their number, the environmentalists were left in no doubt as to the likely fate of the valley if the industrialists got their way.

While the implications of all of this were sinking in,

Intermarine was proceeding to negotiate yet another regulatory hurdle. On 18 March 1979 it formally applied to Mayor Baudone of Sarzana for a building permit concerning certain works essential to the construction of the minehunters. The proposed alterations comprised a work station in the construction hall for laying up each new hull (excavated out of the ground to ensure that the vessels would fit under the roof of the hall), a basin to accommodate the vessels during outfitting and an extension to the existing office building. Although it was not made clear in the application, a new canal linking the work station to the river was also necessary, 200 m long and 10 m wide. In arguing the application, Intermarine reiterated its obligation to deliver the minehunters by 1985 and again emphasized the employment opportunities which the order had created. Here the company wielded both carrot and stick: while the prospect of "250 or more" jobs was now being ventured, Baudone was reminded that 90 of the 150 employees were temporarily laid off due to a lack of orders at the shipyard and were receiving welfare payments from the commune. To make absolutely sure that there was no misunderstanding about the situation, the letter added that "the return to work by all personnel, foreseen for 16 April, is dependent on carrying out the plan discussed above". In the face of such impeccable reasoning, Mayor Baudone duly authorized the alterations.

Now, however, the alarm had been well and truly sounded, and a broad coalition began to emerge in opposition to Intermarine's plans. The most prestigious of the critics was the Italian chapter of the World Wildlife Fund (WWF). In a lengthy assessment of the issue circulated to 13 central, regional, provincial and local authorities, WWF argued that modifying the Columbiera Bridge was the key to opening up the valley to industrial development. The reasons for resisting such a step were many: the exceptional beauty of the valley's landscape; the suitability of the area for agriculture, tourism and small-scale craft industries; the value of the river as a source of drinking water to many communes; the risk to coastal water quality; and the threat to the beaches of Massa-Carrara should the flow of the Magra be affected by new development. The fact that many of these objections were linked to the area's flourishing tourist industry only served to strengthen the opposition from those local commercial interests which were dependent on the large numbers of summer visitors for their livelihood. The Provincial Bureau for Tourism in Massa-Carrara, for example, pleaded with Intermarine to consider the implications of its plans and to delay a final decision on modifying the bridge until all interested parties had made their views known. The Massa-Carrara branch of the Italian Angling and Subaqua Federation responded with an unconditional rejection of the proposals on the grounds that the Magra, so far spared serious pollution, should not be turned into an open sewer. And the Carrara Traders Association objected strongly to the opening of the bridge and to any further extraction of sand and gravel from the river.

News of Intermarine's plans also arrived in Rome. To be precise, it reached the ears of Falco Accame, an ex-admiral who was now a member of parliament for the Communist Party and chairman of the parliamentary committee on defence. Retaining a critical interest in naval affairs, Accame took up the matter and from that moment on became a one-man parliamentary agitator on the issue, persistently asking pertinent questions on the many alleged irregularities surrounding the order and instigating a parliamentary inquiry into the general issue of arms supplies and the granting of defence contracts. He was not alone in his activism. Intermarine's announcement was also a signal to Luigi Biso of Italia Nostra that a more adversarial approach was called for. Seeing the need for decisive action he switched to a campaigning style, making good use of his influential position in the local community. His response to the announcement was to emphasize the longer-term implications of the proposals, likening the opening of the bridge to the Trojan Horse which would secure the entry of the industrialists into the valley beyond. But he did not let this new activism affect his political judgment, for he then made what turned out to be a very astute proposal: Italia Nostra would agree to a modified bridge, but only if the valley were designated a protected area under Liguria's Act no. 40 of 12 September 1977, which provided for the establishment of 13 "regional parks". Such a strategy would accommodate the interests of Intermarine and at the same time secure the long-term conservation of the Magra valley.

But for the moment this proposal fell on stony ground. After all, Intermarine was already assured of the active support of local industrial interests and of the trade unions and had noted the passive assent of the commune to its development. It had also been granted permission to expand its construction facilities and office accommodation and it possessed a signed agreement with the roads authority approving the necessary modifications to the bridge. Moreover, for the first time in its history, Intermarine was operating at profit – some L 380 million ($900 000) in 1979. To be sure, this turnaround in the company's finances was in no small measure due to the first payment on the contract from the Ministry of Defence, equivalent to 25 per cent of the minehunter contract sum. Further, the positive effect of this income on Intermarine's bank balance was magnified by the decision to postpone the start of construction work on the first of the ships until 1980, when the work station, access canal and outfitting basin would be completed. This, however, left the shipyard, with no experience of building 500 tonne vessels, little more than a year in which to produce the first example.

Given the novelty of the design, it was almost inevitable that many unforeseen technical difficulties would arise. Nobody was therefore surprised when a series of problems were encountered in developing the propulsion, electrical, safety and logistical systems of the vessels and in ensuring that certain components were non-magnetic. On top of that, several strikes, of both Intermarine's own workforce and that of its suppliers, delayed progress even further, and it soon became obvious to both Intermarine and the Ministry of Defence that the first delivery deadline would not be met. Indeed, so intractable were the problems that Intermarine's incessant requests for deferments to the delivery dates finally added up to a total of 604 days. The Ministry, however, was not altogether impressed by some of the arguments put forward to justify these delays, and eventually consented to extensions totalling 357 days, giving a new delivery date for the first of the minehunters of 4 May 1982.

An important consequence of these technical and managerial difficulties was a substantial increase in the cost of the vessels. In 1980, just two years after the contract was signed, the price of the four minehunters had more than doubled – from L 64 billion to L 140 billion (then equivalent to $330 million). Despite the problems and the growing controversy surrounding the Columbiera Bridge, Intermarine remained incurably optimistic. A clear indication of its optimism was to be found in the determined sales efforts undertaken at this time. Intermarine believed that a sizable market existed for its minehunter design, especially amongst those Third World countries anxious to upgrade their naval forces. Being newcomers to the defence business, the only way to set up an effective sales operation in a short space of time was to employ freelance salesmen who already enjoyed an extensive network of contacts. One of these agents was Gino Birindelli, formerly an admiral with the Italian navy and NATO chief of southern European waters. On retiring, Birindelli began operating privately as a sales representative for a number of shipyards, receiving a healthy commission for any orders he secured. One of the many potential customers approached was the government of Malaysia, which was keen to improve its defences against mines. Contact was duly established, negotiations proved to be fruitful, and on 5 January 1981 a contract was signed between Intermarine and the Malaysian government for four minehunters. It was the second major order for Intermarine, and a development which reinforced the company's position in its dispute over the Columbiera Bridge.

There remained one slight snag: Birindelli did not receive any commission for the deal. Aggrieved at this omission, the former admiral brought a court action against Intermarine claiming breach of contract. He alleged that the company terminated its contract with him in December 1980 when the deal had been set up, only to sign the agreement with Malaysia a few days later. Intermarine claimed that Birindelli's contract simply expired, and since no deal had been agreed by then there was no legal obligation to pay him any commission. Not surprisingly, the case generated considerable negative publicity for Intermarine, and the shipyard finally decided to cut its losses and settled the claim out of court.

Meanwhile, the prospects for the delivery of the minehunters to the Italian navy remained as bleak as ever. Even if Intermarine encountered no further technical problems or industrial disputes, there was still no guarantee that it would be permitted to make the necessary modifications to the Columbiera Bridge. Confronted with such a desperate situation, the company elected to take a bold and decidedly novel step: in a letter to the Ministry of Defence dated 7 May 1981, it asserted that the minehunters would be delivered even if the bridge was not modified. The solution was surprisingly simple. Each vessel could be lifted onto a transporter and towed some 390 m around the Columbiera Bridge, there to be returned to the river. In discussing the feasibility of such an operation, Intermarine pointed out that ships exceeding 1000 tonnes had been transported over land during the Second World War; bearing in mind the strength of the minehunter design and the technical advances in transportation since then, no serious problems were to be expected. Indeed, the company attached a declaration from the Fibreglass Construction Department of the navy's General Arsenal Office in La Spézia confirming the technical feasibility of the scheme.

This alternative method of getting the minehunters to the sea, backed up with a technical assessment, inevitably attracted much attention when it became known, as it effectively offered a practical alternative to opening the Columbiera Bridge. This logical conclusion was apparently not shared by Intermarine itself, for the company was later to claim that, although technically feasible, it was never considered as a realistic option; deep basins would need to be dredged at the edge of the river where the vessels were lifted out and returned to the water, and the bridge would have to be closed to road traffic for one or two days. The proposal was made only to reassure the Ministry of Defence at a time when pressure was building up to find a solution to the problem of the bridge. Allegedly, an estimate of the likely costs of the operation was not even drawn up (although a later study by the University of Pisa suggested that transporting the vessels over land would cost only a third of the amount involved in modifying the bridge). Be that as it may, Intermarine was later to discover that others did not take the idea quite so lightly.

By now the original delivery date for the first minehunter – 13 May 1981 – had passed quietly by, and still the Columbiera Bridge serenely spanned the clear waters of the river Magra. No solution was in sight and the opposition was gathering strength. But worse was to come. First Intermarine learned that the public prosecutor in Genoa had started a judicial investigation into the role of De Bernardis in signing the 1976 agrement between ANAS and the shipyard – examining whether there was any "mixing of private and public interests" as it was quaintly put. Then, just to be on the safe side, ANAS wrote to Mayor Ennio Silvestri of Amèglia, requesting that the commune refrain from allowing Intermarine to start work on the bridge without its express permission. That it had no intention of giving such approval shortly became clear.

On 27 January 1982 ANAS announced that it was revoking the 1976 agreement which authorized Intermarine to replace one of the fixed spans of the Columbiera Bridge with a movable section. If its reasoning for taking this action was convincing, the legality of the move was less so as the original agreement had yet to be declared invalid. The only way around this obstacle was for ANAS, together with the Minister of Public Works, Franco Nicolazzi, to go through the cumbersome process of issuing a decree specifically for the purpose of legitimizing the decision. With no other means of remedying the situation, this was indeed done. The decree was signed on 3 March and stated that there were "valid motives for revoking the concession", namely that the public interest served by opening the bridge was now less than in 1976 (though in what way was not elaborated), and that the substantial increase in traffic using the bridge in recent years meant that major works to the structure would inevitably cause serious disruption.

But even this apparently immutable step was not quite what it seemed, for within days ANAS decided, under pressure from industrial interests, to review the situation once more. The new strategy aimed at securing consensus amongst the various interests. With this in mind, Nicolazzi invited many of the parties to a meeting in Rome on 23 April. Here he presented a compromise proposal: to modify the bridge as suggested by Intermarine but to limit its use to three or four times a year, only at night and only for vessels sailing downstream. This arrangement had the merit that each minehunter could depart from the shipyard as it was completed while at the same time ostensibly preventing the Magra from being converted into a major shipping channel. Nevertheless, the proposal met with objections from both sides. The compromise did not satisfy the industrialists' demand that the valley be recognized as the "logical" site for industrial expansion out of La Spézia and therefore be opened up to shipbuilding and engineering works. The environmentalists, rejected any moves to modify the bridge, seeing it as the thin end of a very substantial wedge. But what angered them most was that many of the interests opposed to industrial development, such as the commercial organizations in Massa-Carrara, were not even invited to the meeting. This lack of agreement left Nicolazzi in something of a quandary. He nevertheless came to the remarkable conclusion that it had in fact been inopportune to revoke the 1976 agreement and that such a move represented an excessive use of central government power. The only course of action open to him to repair this error of judgment was somewhat embarassing – another decree would need to be issued to revoke the decree that revoked the agreement. An order to this effect was consequently drawn up and duly signed into law on 5 October 1982.

Nicolazzi's failure to invite representatives from Massa-Carrara to the meeting turned out to be a further political error, for the risks posed to the area's beaches were attracting increasing attention. Nicolazzi himself had received a letter on the problem only the month before from the National Committee for the Protection of Italian Soil and Beaches. He was warned that the flow of the Magra had changed markedly due to the extensive excavations along its lower reaches and that any further work would pose a serious risk to the stability of the Massa-Carrara beaches. He was further reminded that only two years previously his own ministry had spent L 3 billion ($7 million) on sea defence measures in order to protect this very coastline. The eventual consequences for the local beaches of opening the Columbiera Bridge were also a central issue at a conference on tourism organized shortly afterwards by the commune of Massa. It was argued that attempts to protect the coastline would have no effect if industry were to be allowed to locate along the Magra since the inevitable changes to the profile of the estuary would almost certainly affect the existing sediment depositional patterns and lead to a gradual erosion of the Massa beaches.

Despite all this activity, it had not escaped the notice

of the Ministry of Defence that the revised deadline for the delivery of the first minehunter – 4 May 1982 – had come and gone with no sign of any ship. In fact not only was the first vessel incomplete, but there was virtually no prospect of it being delivered that year – even if by some miracle the Columbiera Bridge issue were to be speedily resolved. The repercussions of withdrawing from the contract were nevertheless of sufficient proportion that the Ministry was persuaded to give Intermarine one last chance. A final, final deadline was fixed for 15 January 1983.

CHARTING A NEW COURSE

While the Minister of Public Works was trying to piece together a compromise agreement on the Columbiera Bridge, a more ambitious plan was coming to fruition in the provincial administration of La Spézia. The sponsor of this alternative proposal was none other than the president of the province, Sauro Baruzzo. Best described as a political realist, Baruzzo was not enamoured with the idea of developing the Magra valley, and he would have preferred Intermarine to move its construction facilities to La Spézia where yards were now closing due to a lack of orders. (Italian shipbuilders had been operating at about half capacity for the previous three years because of the world recession in shipping.) But given the local economic value of the company, he accepted the need to open the bridge in order to allow the minehunters through to the sea. His own proposal was very much a political compromise and broadly corresponded to Luigi Biso's earlier suggestion: that, as a precondition of the commune of Amèglia granting approval to modify the bridge, Liguria should designate the Magra valley as a protected area.

This idea had already been the subject of a report by a special committee set up by Liguria in 1980 and had been approved by Amèglia in 1981. Further exploratory discussions with the other communes along the Magra and with Intermarine and other local industrial interests also proved to be encouraging – so encouraging, in fact, that in 1982 Baruzzo managed to persuade Liguria to introduce a bill to establish a so-called "river park" in the Magra valley. Drafting such a bill was no easy matter: both industry and the many communes along the valley were concerned that the controls should not rule out new development, yet the prime purpose of the river park was clearly to conserve the valley. The bill was nevertheless drafted, debated and approved by the regional council by 24 August of that year – a remarkable achievement, suggesting that the river park was an idea whose time had come. Even so, the original draft of the bill was rejected by the Minister of Regional Affairs as being too accommodating to industrial development under the provisions of Liguria's own Natural Environment Protection Act of 1977.

The Establishment of the Magra River Park Act, was finally approved on 19 November 1982. It could hardly be described as an elegant piece of legislation, combining as it did general provisions relating to the establishment and management of the river park with a number of ad hoc measures designed to regulate undertakings already located in the valley, particularly Intermarine. The main features of the legislation were as follows:

- A river park was to be established in the 16 communes along the Magra and Vara rivers in the region of Liguria. The term "river park" is highly apt as the boundary of the protected area was to extend just 150 m from each river bank, giving a total area of 20.5 km^2. This is not quite as curious as it seems, for the steep slopes of the valley sides served to concentrate disruptive activities along the narrow banks of the river.

- The establishment of the park had three broad objectives:
  - to improve protection against all forms of degradation;
  - to protect, restore and value the characteristic qualities of both the natural environment and local agriculture;
  - to encourage recreation, agriculture, fishing and tourism within the context of the first two objectives.
- A river park management authority was to be set up. This was to comprise: first, an assembly of representatives from the 16 communes and the province of La Spézia; second, a management committee and president elected by the assembly; and third, a scientific committee appointed by the assembly to provide technical support.
- The management authority was to draw up a plan for the park to provide the context for future development.
- Special development control procedures were to apply. All building permits and concessions which had already been granted and which were not compatible with the objectives of the river park were no longer to be valid unless construction commenced within 30 months of the legislation coming into effect. Improvements to existing facilities in the interests of hygiene, safety or working conditions were to be permitted until 31 December 1983.
- Special rules were to apply to existing undertakings which were incompatible with the objectives of the park. These provisions were simple in their phraseology but far-reaching in their potential effect: all industrial undertakings which were incompatible with the objectives of the river park were to be relocated outside the protected area. This measure was clearly aimed at the special case of Intermarine, and effectively killed the company's long-term plans.

The designation of the Magra valley as a regional park was without doubt a substantial political achievement, given the diversity of the interests involved and the jealously guarded autonomy in development planning enjoyed by the communes. What the measure failed to do was to pacify the opponents of industrial development. In fact, paradoxically, the scheme received more support from industry and the trade unions than from environmentalists. To be sure, it was a leading environmentalist, Luigi Biso, who had originally floated the idea of such a deal, but the legislation as finally approved suffered from four crucial shortcomings from the point of view of the conservationists. First, the park offered only a limited degree of environmental protection: the protected area extended to 150 m on either side of the river instead of the 200 m which the environmentalists regarded as necessary. The communes were still allowed considerable scope for approving new development along the river, and the transitional period permitted the boatyards to continue improving their facilities, even though the river park legislation inferred that they be relocated away from the valley.

Second, the degree of protection offered by the river park would only become clear once the detailed land use plan had been drawn up. But given the dominant role of the communes in the management authority, the environmental groups were by no means convinced that a sufficiently restrictive plan would emerge.

Third, to be effective a plan has to be implemented. By now, however, the opponents of industrial development had become somewhat cynical about the resolve of the local authorities in enforcing their own legislation. The illegal excavation of sand and gravel, the siting of boatyards and the construction of Intermarine's premises all served as forcible reminders that those charged with implementing policy often meet with fierce resistance, and that a weak public authority may well capitulate before the determined action of a powerful private interest.

Finally, the river park, being the artefact of the regional administration of Liguria, terminated abruptly at the border with Tuscany. As 44 km of the Magra was situated in this adjoining region, a large part of the river remained outside the scope of the new controls. Tuscany, although strongly opposed to industrial development along the lower stretch of the Magra and therefore to any move to modify the Columbiera Bridge, was by no means keen to establish a corresponding protected area along the upper reaches of the river, as this would have serious consequences for the local marble industry which quarried huge amounts of the rock from the Apennines and was one of the region's biggest employers.

To keep up its pressure on the river park authority so as to ensure that the river park legislation be effectively applied, it was essential that the environmental coalition be seen to enjoy extensive public support. Fortuitously, a good indication of this support was provided during the drafting of the legislation. Well aware of the groundswell of opinion in its favour, the supporters of the river park organized a protest petition under the unequivocal declaration:

> *No* to the drawbridge that opens the doors to a wild and polluting industrialization of the Magra and our valleys.
>
> *Yes* to setting up a park in Tuscany and Liguria along the banks of the Magra. We demand jobs in agriculture, tourism, fishing, flower-growing and small-scale craft industries.

In August, when the campaign closed, more than 6000 signatures had been collected. The support of several well-known personalities, such as the writer Mario Soldati, added to the publicity value of the venture. This was further increased when the President of Italy, Alessandro Pertini, agreed to receive the petition. If he ever took the trouble to glance through the 300-odd pages, he would doubtless have blinked when he saw one of the signatures

– that of a certain Giovanni Spadolini, occupation: Professor, resident of Florence. Signore Spadolini was in fact rather better known for his second occupation: prime minister.

Also hard at work on two fronts was Luigi Biso. Apart from his involvement in organizing the petition, Biso was taking Intermarine on its word and was busy following up the claim that the Columbiera Bridge need not necessarily be partially demolished in order to get the minehunters to the sea. Intrigued by the possibility of transporting the ships around the bridge, Biso was busy consulting several haulage contractors on the feasibility of the idea. Six firms responded to his inquiry. Remarkably, all insisted that it was a viable option: Angelo Carlotto and Co. and SAIMA simply affirmed that transporting the ships around the bridge was possible; Fagioli telexed to say that it could provide a vehicle capable of carrying up to 680 tonnes; Peyrani claimed that it had transported larger and heavier loads for other clients; and Cometto Trading had calculated that two vehicles would be necessary. Grandi Sollevanenti offered an alternative – and dramatic – solution to the problem: lifting the ships over the bridge with a giant crane. The company maintained that its crane could lift up to 1200 tonnes and submitted sketches of how it proposed to carry out the operation.

It did not take Intermarine long to realize that it had inadvertently opened a can of worms. Instead of simply reassuring the Ministry of Defence that a drastic alternative was available to get the minehunters to the sea, should permission to open the bridge be witheld, they had started everyone wondering whether there might not, after all, be a simple solution to the problem. In fact, inspired by these proposals, all sorts of novel ideas were being hatched. The possibility of ballasting the vessels to enable them to pass under the bridge was much discussed, but the unitary construction of the ships prevented this: the only way to get the ballast into and out of the hull would be to cut off the superstructure, and no facilities existed on the seaward side of the bridge to repair the damage. (Indeed, if it had been

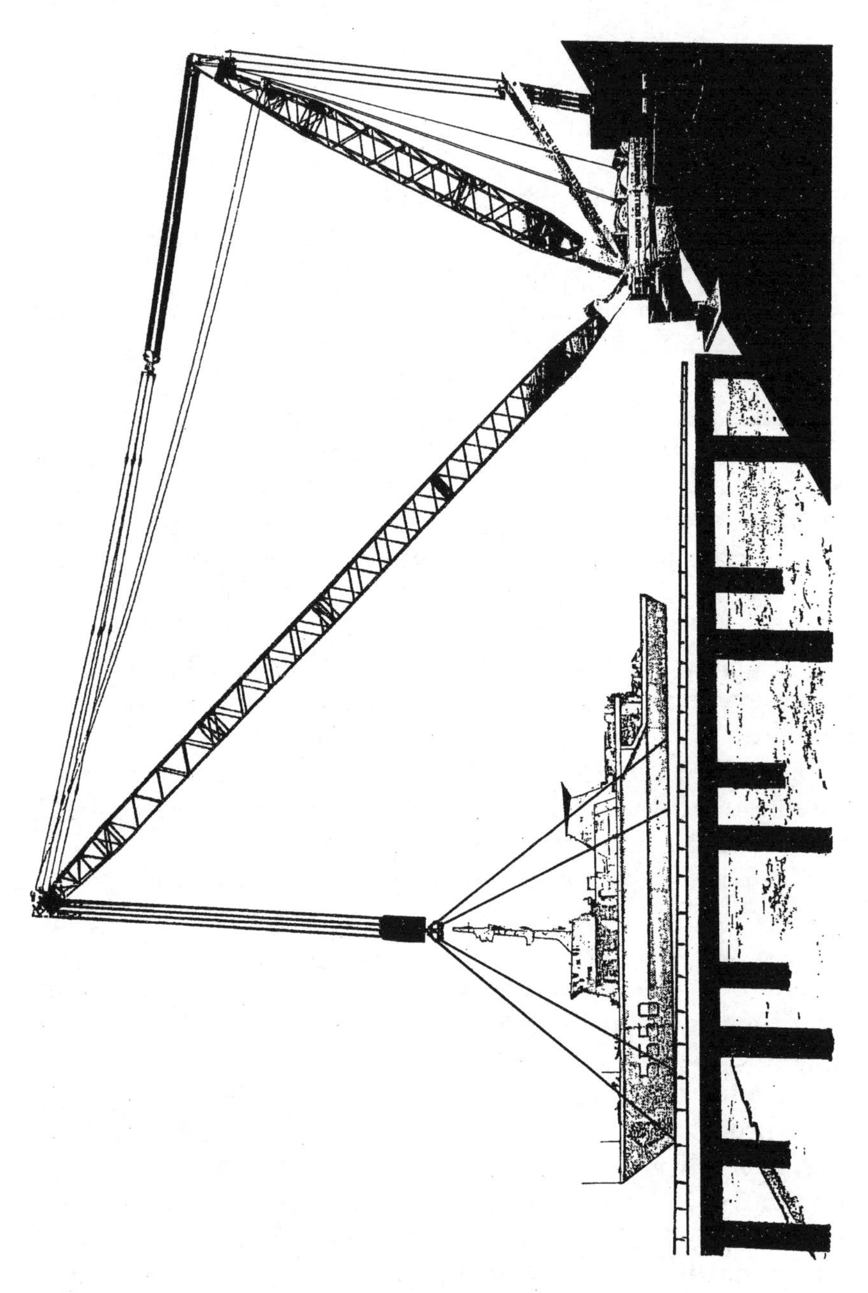

The proposal by Grandi Sollevamenti to lift the minehunters over the Columbiera Bridge

possible to fit the superstructure later, no ballast would have been necessary since the hull itself was low enough to pass under the bridge.) Excavating a canal around the bridge along which the minehunters might be navigated was also suggested, as was the rather more radical solution of using an airship to airlift them to the sea.

To Intermarine, with its eyes set to the future, things were taking a turn for the worse: a decision to leave the Columbiera Bridge intact would be a serious encumbrance to the company's long-term development plans. Action was clearly needed. Accordingly, in November 1982, Intermarine invited a delegation from Amèglia to its offices, under the leadership of Mayor Silvestri, in order to discuss the question further and hopefully to debunk the alternative proposals for getting the minehunters to the sea. It had marshalled its arguments well: transport over land was now ruled out for technical reasons (and a letter from the haulage company SAIMA dated 15 September 1982 was cited as evidence); building the minehunters in component form for assembly at a coastal site was impossible because of the need to lay up the entire structure in a single mould; and ballasting the ships so that they might pass under the bridge was prevented not only by the internal structure of the hull but also by the fact that the 900 tonnes of ballast necessary to ensure clearance would lower the hull to such an extent that it would ship water. The overall impression given by Intermarine was very convincing. Certainly it seemed to persuade the delegation that opening the bridge was the only feasible solution to the minehunter problem, although it later transpired that SAIMA was less than happy with this conclusion. Writing to Luigi Biso on 13 January 1983, it explained that Intermarine had misinterpreted its letter of 15 September: although there were indeed certain problems in transporting the ships around the bridge, SAIMA could certainly carry out the operation provided that the route was properly surfaced, and that loading and unloading were possible (which it believed to be so).

NO EXIT

With the river park legislation now on the statute book and Intermarine insisting that the only means of getting the minehunters to the sea was the old-fashioned way of sailing them there, an agreement on modifying the Columbiera Bridge began to look increasingly likely. This prospect effectively focused local political debate on the issue and increased pressure on Amèglia and ANAS to decide whether the bridge could be modified. The commune finally bit the bullet in January 1983. First to grant permission was the building committee. But the approval of the full council was a far more complex matter, as virtually all the parties had long been internally divided on the question. Indeed, with the bridge now firmly on the political agenda, agreement within the parties seemed further away than ever: the Christian Democrats were mostly opposed to opening the bridge, the Socialists were sharply divided, and the Social Democrats remained undecided. The decisive session of the council was held on 23 January 1983. A heated debate took place, in which the arguments for and against opening the bridge were once more exchanged, but the proposal was finally carried by a vote of 11–1.

One immediate victim of the conflict was the Social Democrat Deputy for Education, Industry and Agriculture, Giorgio Pellegri. Opposed to opening the bridge, Pellegri was so angered by support for the move within the provincial branch of his party that he resigned from his post. All that remained was for Mayor Silvestri to put his signature to the formal authorization, and this he duly did.

Ameglia's vote to approve modifications to the Columbiera Bridge was greeted with a sigh of relief by Intermarine. But no sooner had that problem been overcome than a new threat appeared. On 7 February the Ministry of Defence wrote a five-page letter to Intermarine setting out the troubled history of the minehunter order. Recalling that the delivery date for the first vessel had been progressively deferred from 13 May 1981 to 15 January 1983 and that even this final deadline had not been met (the first minehunter was by then in an "advanced state of preparation"), the ministry informed the company that it now proposed to annul the contract on the grounds that Intermarine had failed to fulfil its obligations. The letter refrained from making a formal announcement to this effect, explaining instead that the appropriate legal procedures for annulling the contract were being investigated. As it happened, the first minehunter was completed shortly after the letter was dispatched. Moored alongside the shipyard, gazing forlornly past the Columbiera Bridge to the open sea, it did not, of course, help the Ministry of Defence very much, and the navy was not amused at seeing the first of its shiny new minehunters rendered impotent by a simple bridge.

And the issue continued to generate scandal. Now it was alleged that Intermarine was busy offering selected people a comfortable job at the shipyard. First, the March 1983 edition of *Parlamento e Forze Armate (Parliament and Armed Forces)*, a news sheet of the parliamentary representatives of the Communist party, published a series of questions to the Minister of Defence, Lelio Lagorio, concerning the delays in the construction of the minehunters. The final question pointedly asked how such delays were possible given "the presence of ex-officers from the Customs Service and even from the navy itself in the management of Intermarine". Then, on 9 April, the newspaper *Il Secolo XIX* reported that an official of the trade union UIL had also joined Intermarine. Robert Quber, secretary with the union, had taken up the position of contracts adviser, giving up his union membership in the process. Critics linked this move with UIL's long support for Intermarine, but Quber denied the allegation.

During the spring and summer of 1983 the pressure on ANAS to finally agree to open the bridge became intense. Michael Trimming, Intermarine's technical director, claimed that potential customers for the minehunters were losing interest because of the problems: Indonesia was now talking about purchasing just two vessels instead of nine, and the US navy, originally planning to buy 30, was now evaluating the possibility of an American yard building the minehunters under licence. Baruzzo reiterated his compromise proposal that ANAS should permit the bridge to be modified and that Intermarine should move its works to La Spezia. The unions again reminded everyone that the minehunter project offered substantial employment prospects in a depressed sector – 400 employees plus 1000 indirect jobs in component and service industries. In August, despite these claims, ANAS informed Intermarine that it was still not prepared to make a decision.

By now the second vessel for the Italian navy and the first for Malaysia had been completed and were moored disconsolately in front of the shipyard to keep the original minehunter company. The ships might not have been operational, but queuing up behind the bridge they quickly became a popular tourist attraction. ANAS nevertheless remained singularly unmoved by the sight, and its continuing intransigence was the signal for Intermarine to play its final card – unless a positive decision on the Columbiera Bridge was forthcoming, the company was faced with no other choice but to start laying off its workers on 3 October.

But other, more significant developments were afoot. Coincidentally, August 1983 saw the inauguration of a new Italian government, replacing the coalition of Giovanni Spadolini. The new team was led by the socialist Bettino Craxi, who hardly had time to warm the seat of his new chair before the affairs of state jostling for attention in Italy's highest office were pushed rudely to one side by matters of a more parochial nature. The instigator of this inconsiderate intrusion was Franco Nicolazzi, who had retained the post of Minister of Public Works in the new cabinet. As the political boss of ANAS, Nicolazzi felt that the time had finally come to rid the country of one of its longest-running public jokes. To this end he once again proposed that the modifications to the Columbiera Bridge be approved so as to allow the minehunters access to the sea. But at the same time restrictions would be imposed on the operating regime of the movable span in order to diminish the attractiveness of the Magra valley as a potential site for further industrial development. In fact the scheme showed a remarkable resemblance to Nicolazzi's abortive proposal of the previous year. Now, however, he was banking on two new elements. First, with the minehunters lining up outside the shipyard, Intermarine was clearly desperate to get the bridge opened – under almost any conditions. Second, it was now suggested that an agreement be drawn up and signed by the government rather than ANAS, thereby lending considerable political weight to the plan.

The first step in this grand strategy was to hold a meeting of the five ministers involved in the issue: Nicolazzi (Public Works), Renato Altissimo (Industry), De Michelis (Employment), Biondi (Ecology) and Spadolini (now, ironically, after signing the petition against opening the bridge, Minister of Defence). The meeting was arranged for 16 November and ended in a certain measure of agreement on the desired strategy, although Alfredo Biondi took a dissenting view on some points, setting out his own standpoint in a short paper to Prime Minister Craxi. While stressing that opening the bridge would be in conflict with both environmental protection and the objectives of the river park, he nevertheless accepted that a "positive solution" was necessary. If other alternatives were not feasible (and the inference was that they were not) then the bridge would have to be modified. This, however, should only be done under three circumstances:

- that existing industries which were incompatible with the objectives of the river park relocate outside the park within three years;
- that a special commission be established to control the operation of the bridge;
- that a satisfactory plan for the river park be drawn up.

Discussions continued at a further meeting on 25 November. It was the compromises hammered out at this second gathering which were to provide the foundation for two key documents: a political compact drawn up under the authority of Bettino Craxi and a draft contract between ANAS and Intermarine. Craxi's compact – sent for signature to the Ministers of Public Works, Foreign Affairs, Industry, Defence, Merchant Navy, Employment, Culture and Ecology and also to the departments within these ministries most involved in the issue and to the President of the province of La Spézia, Sauro Baruzzo – was an astute means of committing the major political protagonists to a common line. It declared that all proposed alternatives to modifying the bridge had been found to be technically impossible and that agreement had been reached on seven points:

- all the relevant authorities would have to give their permission before Intermarine could modify the bridge;
- opening the bridge had now become imperative both in the national interest and for international political reasons;
- the public interest (in the form of traffic movement) and Intermarine's private interest were reconcilable;
- the Ministry of Culture and the regional authority would ensure that environmental interests were protected;
- to ensure improved environmental protection, the Ministry of Public Works would establish a committee for supervising the operation of the modified bridge, comprising representatives from the regions of Liguria and Tuscany, the province of La Spézia, the Ministries of Ecology and Culture, the Geological Institute of the University of Pisa, the Institute of Hydrology of the University of Florence and the president of the river park;
- the appropriate authorities would help to secure the

relocation of Intermarine;

- the appropriate bodies were to draw up a detailed management plan for the river park.

The closing section of the compact explained that Craxi would advise ANAS on the drafting of a contract that was to be appended to the original 1976 agreement with Intermarine. The contract that emerged, signed by ANAS and Intermarine, set out the conditions under which the bridge could be modified. These included:

- that approval for both the contract and the work was to be granted by Liguria, Amèglia, the utilities and the harbour authority;
- that Intermarine would bear the costs of the work (including the costs of diverting traffic and rerouting the gas, electricity and water mains and the telephone cables);
- that Intermarine would operate and maintain the movable section of the bridge, although this would become the property of ANAS;
- that an annual fee of L 20 million ($40 000) would be paid to ANAS by Intermarine plus an amount to be determined each time the bridge was opened;
- that the bridge was not to be opened between 20 December and 10 January, during the month of August or on public holidays, and that traffic was not to be disrupted more than ten times each year;
- that Intermarine pay ANAS L 2 million ($4000) to cover the authority's inspection costs, deposit a sum of L 400 million ($750 000), repayable six months after the modifications were approved, to cover any claims for compensation which might arise as a result of damage caused by the work, and make a further deposit of L 100 million ($190 000) for the duration of the contract;
- that the contract be valid for nine years, but be annulled should Intermarine fail to fulfil its obligations.

By the middle of February, negotiations with the utilities on relocating the telephone cables and the electricity, gas and water mains on the river bed had been completed. And despite the continuing protests of the environmental coalition and the fact that the exact conditions under which the bridge could be opened had yet to be finalized, the contract was signed on 16 February. Seven years after the original agreement with ANAS, Intermarine could at last lay its hands on the Columbiera Bridge.

But if Intermarine was on the defensive over the question of the minehunters contract, it was at long last leading a major assault on the environmentalists. After years of conflict Rocco Canelli's patience was wearing thin, and he had been particularly irked by a telex sent jointly by several environmental groups to the previous Minister of Defence, Mayor Silvestri of Amèglia and the Commander of the Alto Timeno area of the Mediterranean, Admiral Vittorio Gioncada. Sent on 5 October 1981, the telex cast doubt upon the way in which Intermarine had secured the original agreement with ANAS and the way in which the minehunter contract had been secured. It then went on to explain that the shipyard was now assembling the moulded sections of the first minehunter, but since it was possible to do this on the seaward side of the Columbiera Bridge the clear intention was to present the authorities with a *fait accompli*. The minister was therefore asked to intervene before it was too late. Seven groups put their names to the telex: Italia Nostra, WWF, Comitato Difesa e Sviluppo Val di Magra, Rosa Verde, Movimento Cristiano Lavoratori and Società Pescasportivi Val di Magra. Intermarine, in the person of Rocco Canelli, duly brought an action for defamation against 10 representatives of the organizations, including Luigi Biso of Italia Nostra and Roberto Lasagna of WWF. Although the local magistrate had taken up the accusation for consideration in October 1981, his judicial investigation had been so protracted that the case only came to court on 7 March 1984.

Intermarine's accusation was based on three assertions: first, that the environmental groups had claimed that the way in which ANAS came to give its original permission was unclear; second, that the way in which the minehunter contract was granted was also unclear; and third, that the reason the vessels could not pass under the bridge was because of Intermarine's insistence on assembling the sections of each vessel in the shipyard. Judgment was passed on 9 April. In his verdict, Judge Federico Sorrentino found that free speech and open criticism were basic rights of the citizen provided that certain conditions were respected, namely that the remarks made were in the public interest, that they were essentially true and that they were objective. His conclusion in this instance was that these conditions had been met. Finding against Intermarine, Sorrentino ordered the company to pay L 200 000 ($400) damages plus L 400 000 ($800) costs to each defendant.

BREAKTHROUGH

Embarassing it might have been, but the judgment now amounted to little more than a distraction for Intermarine since work was finally about to get under way on modifying the Columbiera Bridge. The original plan had been to remove the most westward span of the bridge and fit a lifting section operated by a heavy hydraulic mechanism situated on the right bank of the river. When the new agreement was negotiated, however, ANAS and Intermarine were keen to retain the option of extending the bridge in the event of the Magra being widened at some future date. The reason for keeping this possibility open was that the river narrowed as it approached the bridge and after heavy rainfall was liable to flood its banks – it had twice caused damage to Intermarine's facilities in this way. The flooding was, of course, a natural characteristic of the Magra valley and only became a serious problem once the river banks had been built upon. It was also part of the extensive geomorphological process of erosion and aggradation which was characteristic of the river and which maintained the beaches along the coastline to the south of the estuary – a process which the neighbouring communes were anxious to leave undisturbed.

But the limited capability of the structure to support eccentric loads severely restricted the options available for incorporating an opening section into the bridge. Indeed, the only feasible solution was to replace one of the fixed spans with a sliding rather than a hinged section, a design which necessitated an extremely clumsy opening operation: a special paved section spanning the centre part of the bridge would be raised by a metre on four hydraulic jacks, then fitted with bogies and towed to one side on specially laid rails before the two beams supporting the section could be slid aside to provide a clear passage for shipping. In practice it was found that the entire operation of opening and closing the bridge would take no less than 12 hours – a fact that was to have rather interesting consequences.

Work was scheduled to commence on Thursday 26 April, and ANAS had stipulated that it was to be completed no later than 30 May. As it turned out, even this timetable was to prove controversial. Fully in keeping with the tradition of the issue, it was only later realized that the Tour of Italy cycle race – the famous *Giro d'Italia* – was scheduled to pass over the Columbiera Bridge on 31 May. When Intermarine and the organizers of the race eventually became aware of one another's plans, a meeting was hastily arranged to discuss the feasibility of rerouting the riders in case delays arose during the construction work. This was found to be impossible as any alternative route would have to pass through Sarzana, an extra distance for the riders of some 10 km, and that particular leg of the race was already the longest of the entire tour. There was nothing to do but hope.

Things got off to a bad start. On the day before work was

due to begin, opponents of the agreement between ANAS and Intermarine started a sit-down demonstration on the bridge. So many demonstrators took part that the workmen were prevented from gaining access to the bridge when they arrived the next day and police were called in to break up the protest. After consultations with the authorities, the contractors decided to come back the following day and try again. The demonstrators doggedly held their ground, and only after the police had physically removed the protestors – causing parliamentary deputy Falco Accame to break his ankle – could construction work begin.

Within two weeks the centre span of the bridge had been cut away and lifted clear. After some eight years the way to the sea was at last open, and Intermarine was not planning to take any chances now. At 6.50 in the morning of 11 May 1984 the first of the minehunters, *Lerici*, was piloted through the new gap and on to La Spézia from where the sea trials could begin. It was immediately followed by the second of the vessels for the Italian navy, *Sapri*, and the first two minehunters for Malaysia, *Kinabulu* and *Ledang*. The battle of the Columbiera Bridge had been won.

Whether the war of the Magra valley had also been won was much less certain. The day after the minehunters had passed through the Columbiera Bridge, Sauro Baruzzo made clear his objections to the contract between ANAS and Intermarine, which he saw as conflicting with the river park legislation. Baruzzo's viewpoint was supported in Amèglia where the council, with the exception of the Socialists, passed a motion declaring its dissatisfaction with the contract for making the opening of the bridge possible.

Meanwhile, back at the bridge, high winds were making it impossible to operate the floating crane and the hearts of the *Giro* organisers were starting to beat a little faster. But the winds died down after a few days and no further problems of any consequence arose. On 26 May the new prefabricated section was lifted into place and there was just enough time left to carry out the necessary finishing work before the riders hove into view and sped across the new span to disappear into the hills of Bocca di Magra.

AN OPEN AND SHUT CASE

Throughout the course of the issue, Intermarine's opponents feared two developments above all else: that the Columbiera Bridge would be opened and that the shipyard would prosper. Both came to pass. Yet, paradoxically, it seems that certain consequences of these developments may ensure that industrial pressure on the Magra valley will not become more intense. First, the fitting of a sliding section in the bridge rather than a hinged span meant that the opening operation became a laborious process, taking a total of 12 hours. For a shipyard situated upstream of the bridge, such a technical constraint effectively precludes frequent passages to and from the sea. Yet carrying out sea trials on newly constructed ships requires just that – daily excursions for a period of three months were required in the case of the first minehunter, and shorter but still comprehensive testing programmes were necessary for the subsequent vessels.

Second, with the vessels in production and starting to demonstrate their capabilities, Intermarine became convinced that the likely demand for the minehunter and related designs would justify a greatly increased investment programme. But the expanded construction, outfitting and mooring facilities necessary to support a more ambitious operation could not be accommodated on the existing site alongside the Magra. The only alternative was to establish a second base. The company accordingly acquired a site in La Spézia, where the general decline in the ship-building industry had left its mark on the local facilities, and began construction of a new yard. With a projected cost of some L 10 billion ($16 million), the new shipyard was designed for the production of Intermarine's larger vessels, allowing the works on the Magra to revert to the construction of smaller pleasure and military craft – designs low enough to pass under the Columbiera Bridge.

Progress in drawing up a management plan for the river park was less impressive, however. Serious problems arose in setting up a new management authority and achieving consensus amongst the 16 communes and the province on both the broad principles and detailed provisions of an environmental protection policy for the valley. The prospects for developing a comprehensive management plan and ensuring its effective implementation seemed poor indeed.

Understandably, perhaps, the opponents of industrial development remain cynical about the true worth of the regional park legislation. They have, after all, learned the hard way that well-intentioned plans are not always built on the firmest of foundations, and that in such circumstances a little judicious pressure or a brazen initiative by a powerful private interest can often pre-empt potential opposition. They would also argue that the river park, created primarily as an environmental compensation for the opening of the Columbiera Bridge, is a shakier construction than most of its kind. Whether such a rickety structure can withstand determined assaults on its integrity must be open to question – as indeed was conceded in an open letter concerning the Columbiera Bridge issue published in the magazine *Panorama* on 27 August 1984:

> While the problem is of major environmental significance, it is also characterised by other, financial aspects no less relevant, . . . that a minister of the Republic, presently charged with responsibilities of leadership and coordination, cannot ignore or pretend to ignore.
>
> Alfredo Biondi
> Minister of Ecology, Rome

FURTHER READING

Not all of the issues covered by the case studies have been fully described in the available literature, and where information is available it is sometimes only to be found in specialist papers and journals or in foreign-language publications. Neither broad analyses nor comprehensive and detailed information have been published on the Rhine salt issue, the acid emissions control policy of the European Community or the Magra valley; these chapters were therefore based to a large extent on original research and on some French, Dutch and Italian material.

CHAPTER 2: A SHAGGY FISH STORY

By far the best analysis of the Tellico Dam issue is by Stephen J. Rechichar and Michael R. Fitzgerald, *The Big Dam and the Little Fish: TVA's Tellico dream in an era of intragovernmental regulation*, paper presented to the ASPA National Conference on Public Administration, New York, 16–19 April 1983.

The 1978 judgment by the Supreme Court (TVA v. Hill, 437 US 153 [1978]) sets out the judicial history of the case, the majority opinion of the court and the dissenting opinions.

The implications of the snail darter case for the application of the Endangered Species Act is discussed at length in Eric Erdheim, "The Wake of the Snail Darter: insuring the effectiveness of Section 7 of the Endangered Species Act", *Ecology Law Quarterly*, 9, 1981, 629–682.

A comprehensive description of the snail darter and its habitat is given in David A. Etnier, "*Percina (Imostoma) tanasi*, a new percid fish from the Little Tennessee River, Tennessee", *Proceedings of the Biological Society of Washington*, 88, 1976, 469–488.

The status of the snail darter following its rediscovery and the proposals for its long-term protection are described in *Snail Darter Recovery Plan*, Denver, US Fish and Wildlife Reference Service, 1983.

For a good overview of US political and judicial institutions and how they function see David McKay, *Politics and Power in the USA*, Harmondsworth, Penguin, 1987.

CHAPTER 3: RHINE BRINE

Remarkably, no detailed analysis of the Rhine salt issue is available and only very limited information on the problem is available in English.

A brochure on the history of the problem is available from the IAWR (the international organisation for drinking water companies in the Rhine basin) under the title *Salz im Rhein, Rost im Rohr*, Amsterdam, 1988. It is also published in Dutch as *Zout in de Rijn – roest in de buis*.

The main legal issues involved were reviewed in the papers presented to the conference "Transboundary Pollution and Liability: the case of the river Rhine", held in Rotterdam on 19 October 1990 and organized by the Institute of Environmental Damages of the Erasmus University, Rotterdam.

CHAPTER 4: ACID DROPS

The full reference for the Large Combustion Plant Directive is: "Council Directive on the limitation of emissions of pollutants into the air from large combustion plant", Council Directive 88/609/EEC, *Official Journal of the European Communities*, L 336, 7 December 1988.

The cost-benefit analysis commissioned by the European Commission on the damage costs of acid deposition was published as *Acid Rain: a review of the phenomenon in the EEC and Europe*, London, Graham and Trotman, 1983.

A good review of acidification and the political response to the problem is to be found in John McCormick, *Acid Earth: the global threat of acid pollution*, second edition, London, Earthscan, 1989.

CHAPTER 5: ENDANGERED CULTURE vs ENDANGERED SPECIES

The most comprehensive assessment of the bowhead issue is Nigel Dudley and Joanna Gordon Clark, *Thin Ice*, Cambridge, Marine Action Centre, 1983. Other broad-ranging papers are John Bockstoce, "Battle of the Bowheads", *Natural History*, May, 1980, 52–61; Marion Fischel Johnson and William F. Gusey, "The Bowhead Whale: life history and issues", *Ecolibrium*, 11, 1982, 1–9; and Edward Mitchell and Randall R. Reeves, "The Alaska Bowhead Problem: a commentary", *Arctic*, 33, 1980, 686–723.

The biology of the bowhead is described in Mary K. Nerini et al., *Life History of the Bowhead Whale (Balaena mysticetus)*, Seattle, National Marine Fisheries Service, (undated).

Further information on the Inuit whaling culture can be found in David Boeri, *People of the Ice Whale*, New York, Norton, 1985.

Data on catch rates are published in the annual reports

of the International Whaling Commission, Cambridge (UK). These reports comprise the proceedings of the annual meetings of the IWC and include the scientific assessments which are carried out with a view to setting the annual catch limits.

Published material on the way in which the catch rates are negotiated at the IWC annual meetings is scarce. An interesting account of the 1979 annual meeting is given by James E. Scarff, *Report from London: US vs. the Whales, or what really happened at the IWC meeting?*, Oakland, The Whale Center, 1980.

The problems of assessing historical bowhead populations and catch rates have been dealt with in various publications. A good example is W. Gillies Ross, "The Annual Catch of Greenland (Bowhead) Whales in Waters North of Canada 1719–1915: a preliminary compilation", *Arctic*, 32, 1979, 91–121.

An authoritative review of the International Convention for the Regulation of Whaling is given by Simon Lyster, *International Wildlife Law*, Cambridge, Grotius, 1985. A detailed legal analysis of bowhead whaling forms part of a broader article by Sharon O'Brien, "Undercurrents in International Law: a tale of two treaties", *Canada-United States Law Journal*, 9, 1985, 1–57.

An historical and critical review of whaling and its regulation can be found in Jeremy Cherfas, *The Hunting of the Whale: a tragedy that must end*, London, Bodley Head, 1988.

CHAPTER 6: A CERTAIN ACCIDENT

A comprehensive analysis of the Bhopal disaster is to be found in Paul Shrivastava, *Bhopal: anatomy of a crisis*, Cambridge, Mass., Ballinger, 1987.

Extended coverage of the accident was given in two issues of *Chemical and Engineering News*, dated 11 February 1985 and 2 December 1985.

Union Carbide's assessment of the accident and its contributory causes was published in March 1985 as *Bhopal Methyl Isocyanate Incident Investigation Team Report*, Danbury, Conn., Union Carbide Corporation. A contrasting perspective is given in ICFTU and ICEF, *The Trade Union Report on Bhopal*, International Confederation of Free Trade Unions, 1985.

CHAPTER 7: A BRIDGE TOO LOW

It is both remarkable and unfortunate that no comprehensive study of the Columbiera Bridge issue has been published. The material that is available is largely to be found in the Italian popular press and is almost without exception limited to shorter pieces on specific aspects of the case. A series of articles on the affair was, however, published in *Punto Critico* in the period 31 January–26 March 1986.

The four-volume report of the parliamentary inquiry into arms supplies includes a study of the Intermarine case and reproduces a considerable number of the documents relevant to the issue. It was published as *Relazione della Commissione Parlamentare d'Inchiesta e di Studio sulle Commesse di Armi e Mezzi ad uso Militare e sugli Approvvigionamenti*, Legislatura VIII, Doc. XXIII, 8 June 1983.

A volume on the history and environment of the Magra valley was published in 1981 by the Apuo-Lunense branch of Italia Nostra under the title *Tra Fiumi, Mare e Terraferma*.

INDEX